Public Space

Public Space: Between Reimagination and Occupation examines contemporary public space as a result of intense social production reflecting contradictory trends: the long-lasting effects of the global crisis, manifested in supranational trade-offs between political influence, state power and private ownership; and the appearance of global counter-actors, enabled by the expansion of digital communication and networking technologies and rooted into new participatory cultures, easily growing into mobile cultures of protest.

The highlighted cases from Europe, Asia, Africa and North America reveal the roots of the pre-crisis processes of redistribution of capital and power as an aspect of the transition from the consumerist past into the post-consumerist present, by tracing the slow growth of social discontent that has led only a few years later to the mobilization of a new kind of self-conscious globally-acting class.

This edited volume brings together a broad range of interdisciplinary discussions and approaches, providing sociologists, cultural geographers, and urban planning academics and students with an opportunity to explore the various social, cultural, economic and political factors leading to reappropriation and reimagination of the urban commons in the cities within which we live.

Svetlana Hristova is an urban sociologist, researcher, lecturer and associate professor at the Faculty of Arts of the South-West University in Blagoevgrad, Bulgaria. In 2009 she initiated the working group Urban Management and Cultural Policies of the City at ENCATC, which evolved into a thematic area with the same name. She is the author and editor of numerous publications on urban cultures, public spaces and sustainable development, such as *Culture and Sustainability in European Cities: Imagining Europolis* (2015).

Mariusz Czepczyński is a cultural geographer, and a professor at the Department of Spatial Management, University of Gdańsk, Poland. He is also active in applicative consultancy and advisory work, recently for the mayor of Gdańsk, the Polish Metropolitan Union, the City Hall of Lodz, DS Consulting and PwC. His research has focused on cultural landscapes, post-socialist cities, heritage and urban transformations, and the results have been published in several papers and books, including *Cultural Landscapes of Post-Socialist Cities: Representation of Powers and Needs* (2008).

Design and the Built Environment

Series editor: Matthew Carmona

This series provides a means to disseminate substantive research in urban design and its allied fields. Contributions are welcomed which are the result of original empirical research, scholarly evaluation, reflection on the practice and the process of urban design, critical analysis of particular aspects of the built environment, or important conference proceedings. Volumes should be of international interest, although they may focus on the particular experience and practice in one country. They may reflect theory and practice from across one or more of the spatial scales over which urban design operates.

https://www.routledge.com/planning/series/DESBE

Public Space

Between Reimagination and
Occupation

Edited by Svetlana Hristova and
Mariusz Czepczyński

Routledge
Taylor & Francis Group

LONDON AND NEW YORK

First published 2018
by Routledge
2 Park Square, Milton Park, Abingdon, Oxon OX14 4RN

and by Routledge
711 Third Avenue, New York, NY 10017

Routledge is an imprint of the Taylor & Francis Group, an informa business

© 2018 Svetlana Hristova and Mariusz Czepczyński

The right of the editors to be identified as the authors of the editorial material, and of the authors for their individual chapters, has been asserted in accordance with sections 77 and 78 of the Copyright, Designs and Patents Act 1988.

British Library Cataloguing-in-Publication Data
A catalogue record for this book is available from the British Library

Library of Congress Cataloging-in-Publication Data
Names: Hristova, Svetlana, editor. | Czepczyński, Mariusz, editor.
Title: Public space : between reimagination and occupation / edited by
    Svetlana Hristova and Mariusz Czepczyński.
Description: First Edition. | New York : Routledge, 2018. | Series: Design
    and the built environment | Includes bibliographical references and
    index.
Identifiers: LCCN 2017016947| ISBN 9781472453648 (hardback) |
    ISBN 9781315603018 (ebook)
Subjects: LCSH: Public spaces.
Classification: LCC HT185 .P824 2018 | DDC 307.76—dc23
LC record available at https://lccn.loc.gov/2017016947

ISBN: 978-1-4724-5364-8 (hbk)
ISBN: 978-1-315-60301-8 (ebk)

Typeset in Sabon
by Swales & Willis Ltd, Exeter, Devon, UK

Printed in the United Kingdom
by Henry Ling Limited

Contents

Illustrations

Figures

Tables

Contributors

Nuray Bayraktar has been associate professor at Baskent University, Ankara, Turkey, since 2009 and was at Gazi University in Ankara from 1990 to 2009. She worked as an architect at private offices and companies from 1980 to 1990. Her publications are centered on housing, urbanization and urban design, and she has been published in several books, periodicals and newspapers. She has also won awards in several social and architectural competitions. She is a member of the academic board of the journal of the Chamber of Architects and the director of the 'Architectural Heritage of Housing in Ankara between 1930 and 1980: Research, Documentation and Developing Conservation Criteria Project'.

Mariusz Czepczyński is a cultural geographer, and a professor at the Department of Spatial Management, University of Gdańsk, Poland. He is also active in applicative consultancy and advisory work, recently for the mayor of Gdańsk, the Polish Metropolitan Union, the City Hall of Lodz, DS Consulting and PwC. His research is focused on cultural landscapes, post-socialist cities, heritage and urban transformations, and the results have been published in several papers and books, including *Cultural Landscapes of Post-Socialist Cities: Representation of Powers and Needs* (2008).

Megan Dixon completed her PhD (June 2009) in Urban and Cultural Geography at the University of Oregon in the US. She also has a PhD in Slavic Languages and Literature from the University of Wisconsin-Madison and has taught courses in Russian literature and culture. Her research interests focus on social mixing in urban spaces, landscape change and socio-spatial patterns in post-Soviet Russia. Her research on the Chinese presence in European Russia has been funded by the US National Science Foundation, the US Society of Women Geographers, and the Tokyo Foundation. She is currently an instructor at The College of Idaho in the western US where she studies attitudes to US public lands and the local landscape effects of immigration.

Daniela Dumbraveanu is associate professor in the Department of Human and Economic Geography, University of Bucharest, Romania. Her research interests are related to human geography with a particular focus on tourism (destination identity and regional development). She has written a number of books and articles about the development of tourism in Romania. She also has research interests in the geography of health (particularly the migration of medical personnel).

Wael Salah Fahmi is professor of Urban Design in the Architecture Department of Helwan University, Cairo, Egypt. As a visiting academic at the University of Manchester, Wael has been involved in joint research and publications in various academic journals. His research interest focuses on Greater Cairo's urban growth, housing problems, and gated communities (cities), the rehabilitation of historical districts (Habitat International), the cemetery informal settlements (Arab World Geographer), garbage collectors community, street movements within Cairo's public spaces (Environment and Urbanization), and Cairo's contested European Quarter (International Development Planning Review). He has been published in edited books, in addition to four books and two co-authored books. His latest, forthcoming co-authored book discusses Cairo's planning and housing issues.

Svetlana Hristova is an urban sociologist, researcher, lecturer and associate professor at the Faculty of Arts of the South-West University in Blagoevgrad, Bulgaria. In 2009 she initiated the working group of *Urban Management and Cultural Policies of the City* at ENCATC, which evolved into a thematic area with the same name. She is the author and editor of numerous publications on urban cultures, public spaces and sustainable development, such as *Culture and Sustainability in European Cities: Imagining Europolis* (2015).

Francisco Adolfo García Jerez is professor at the Department of Social Sciences of Universidad del Valle (Cali-Colombia) and a member of the research group GISAP (University Pablo de Olavide, Seville-Spain). He was a visiting professor at Appalachian State University, North Carolina, USA, and at the University of Stirling, UK. He has a PhD in Cultural and Social Anthropology from the University Pablo de Olavide and a Master's in Latin American Studies from the University of Aberdeen, UK. His research has been focused mainly on Urban Ethnography, Social Urban Movements and the Anthropology of the City.

Jerome Krase is a noted sociologist, author and social advocate. Representative books titles include *Self and Community in the City* (1982), *Ethnicity and Machine Politics* (with Charles LaCerra, 1992), *Seeing Cities Change* (2012), and *Race, Class and Gentriifcation in Brooklyn* (with Judith N. De Sena, 2016). Professor Krase also co-edited with Ray Hutchinson, *Ethnic Landscapes in an Urban World* (2007) and Race and Ethnicity in New York City (2004), and authored and co-authored many articles in scholarly publications such as the *Journal of Architecture and Planning Research, Revista Internationale di Sociologia, The Polish Review, Visual Communication, Societes, Diogene, Humanity and Society, Sociology and Social Welfare* and *Urbanities*.

Duncan Light is lecturer in the Department of Tourism and Hospitality, Bournemouth University, UK. He is a cultural geographer with research interests in the politics of tourism, and the relationships between urban landscape, identity and memory. He has published on these topics in a range of journals and is also author of *The Dracula Dilemma: Tourism, Identity and the State in Romania* (2012).

Anastasia Moiseeva, PhD, has an engineering degree in the field of cadastral works and urban management from the Kuzbass State Technical University, Russia. She started her work as an urban planner and designer at the Municipal Centre of Urban

Planning and Land Use, Kemerovo, Russia. She received a MSc in Urbanism at the Delft University of Technology, the Netherlands, with a project on 'Transit Space', focusing on transformation of public spaces in post-socialist new industrial cities in Russia; and in 2013 obtained a PhD degree from the Eindhoven University of Technology. Her research was centered on individual behaviors in public space: how newcomers to a city learn about public locations over time.

Pavel Pospech received his PhD degree in sociology from the Masaryk University and currently works as an assistant professor at the Department of Sociology of the Masaryk University in Brno, Czech Republic. As a postdoctoral scholar, supported by a scholarship from the Republic of Austria, he is also engaged in research at the Department of Sociology, the Technical University of Vienna, Austria. His main research interests are urban public space and the production of social order in cities. He has been engaged in projects, focused on the "new public space" of shopping centres and on incivility in urban space.

Philipp Rode PhD is a landscape architect, a partner in zwoPK Landscape Architecture, and a lecturer and researcher at the University of Natural Resources and Applied Life Sciences, Institutes of Landscape Planning and Landscape Architecture, Vienna, Austria. His research is in the field of urban space and socio-economics, focussing on temporary uses of public space, participation and design.

Remon Rooij is assistant professor of Spatial Planning and Strategy at Delft University of Technology and the curriculum coordinator of the 3-year Delft faculty of Architecture and Built Environment undergraduate program. Remon has strong interests in the relation between mobility and the city, from the perspective of urbanism of networks; urban regeneration strategies and sports inclusive spatial planning. Among his recent publications is *Transformation Strategies for Deprived Neighbourhoods: Intervening in a Complex and Vulnerable Reality* (2012; in Dutch).

Eva Schwab, PhD candidate, holds an MSc in Landscape Architecture and practised as a freelance landscape architect before entering a career in research. She is a lecturer and researcher in the Institute of Landscape Architecture at the University of Natural Resources and Life Sciences in Vienna, Austria. Her main research interest lies within socio-spatial urban research, with a special focus on non-formal space production, the politics of public space production and socio-cultural aspects of open space use. She conducts research both in Europe and Latin America.

Michela Semprebon is an urban sociologist. She holds an MSc in Urban Management from the Erasmus University, Rotterdam, and a PhD in Urban Sociology from the University of Milan-Bicocca. She is currently a post-doctorate research fellow at the University of Milan-Bicocca and teaches at the Politecnico University of Milan. She has been published in Italian and international peer-reviewed journals, such as *Partecipazione e Conflitto*, the *Journal of Urbanism*, the *Journal of International Migration and Integration* and has contributed chapters to various books. Her main research interests relate to local immigration policies, political engagement and urban conflicts, housing and the sociology of housing, urban public space, urban policies.

Harry Timmermans is professor of Urban Planning at the Technical University of Eindhoven. The research activities of his multidisciplinary group concern the development and application of innovative models and ICT tools for urban planning and transportation; innovative ICT apps and smart cities. He has been awarded prestigious grants and won many (best paper) awards with his PhD students. He has co-authored more than 500 publications in urban planning, transportation, artificial intelligence, marketing, and environmental psychology. He is a member of national and international boards and committees, including the Public Transportation Marketing and Fare Policy committee, and committee on smart cities and smart mobility.

Craig Young is Reader in Human Geography in the Division of Geography and Environmental Management at Manchester Metropolitan University, UK. His research interests include a focus on geographies of urban change in post-socialist contexts and the cultural geographies and politics of identity (from the individual to the city and the nation) in the context of post-socialist transformation, particularly in the former Eastern Europe. Research in Romania has focused on issues of identity, memory, commemoration and the nation as played out through particular urban landscapes and public spaces, especially those surviving from the state-socialist period.

Sharon Zukin is a professor of sociology in the field of modern urban life, based in Brooklyn College and the Graduate Center, City University of New York. Among her numerous publications are such influential books as *Naked City: The Death and Life of Authentic Urban Places* (2010); *Point of Purchase: How Shopping Changed American Culture* (Routledge, 2004); *The Cultures of Cities* (1995); *Landscapes of Power: From Detroit to Disney World* (1991) (winner of the C. Wright Mills Award); *Loft Living: Culture and Capital in Urban Change* (1982). She is also the co-editor, with Philip Kasinitz and Xiangming Chen, of *Global Cities, Local Streets: Everyday Diversity from New York to Shanghai* (Routledge, 2016).

Introduction

Svetlana Hristova and Mariusz Czepczyński

The quintessence of a city is always based on the diversity, proximity and coexistence of functions, spaces, structures, and people with their needs, hopes, expectations, and beliefs. This more or less balanced urban melange has been continuously (re)constructed by both spontaneous and organised interactions of urban strangers applying high levels of compromise, co-operation and control. The urban is then a constantly renegotiated compromise between diverse groups of urban users with their different social interests and claims, thus nourishing the resilient diversity in the face of imposed unity as a fundamental part of sustainable urban life. The process of renegotiation between citizens unfolds in public spaces and places, and for this reason they are indispensable for healthy urban life. There has never been a city without public spaces, and the livelier public spaces are, the more urban life can be experienced. This process can also be described as a continuous reoccupation and reimagination of urban spaces which re-establishes the normative, ethical and aesthetic thresholds of shared space in a city.

However, urban compromise and the balance between different groups of interests has been disturbed during the last three decades by the neoliberal hegemony over practically all spheres of social life (Hall 2011). The global financial crisis and the imposed austerity politics affected cities particularly hard. As Jamie Peck put it, 'Cities are. . .where austerity bites' (Peck 2012, 629). The present volume offers a closer critical look into the current transformations of public space in different parts of the world through which fundamental changes of contemporary society have been displayed. Or vice versa: starting from Henri Lefebvre's ideas in the *Production of Space* that 'every society – and hence every mode of production – [. . .] produces its own space' (1991, 31), and since, '*ex hypothesis,* [. . .] the shift from one mode to another must entail the production of a new space' (1991, 46), it is worthwhile to ask ourselves what major changes occurred in society that have led during the last decade to the unprecedented radicalization of public space turned into a premise of new spatial strategies, occupations and spectacular terrorist acts in referential cities of the modern civilization. As the 'urban society' predicted by Lefebvre is already a fact today, our hypothesis is that we are in transition to another social form that was still a 'blind field' when Lefebvre wrote *The Urban Revolution*. This new social being is still being established, and its various aspects and possible dimensions have been highlighted since the 1980s under different names within different theoretical frameworks. Today it is still in a process of illumination.

However, schematically we shall outline the contours of this macro-social transformation in relation to its impact on public space by criss-crossing several main trends which have usually been discussed in diverse scientific and public discourses, while they are deeply intertwined and produce in their interconnectedness a new global order with new social actors challenging and transforming contemporary public space. *Financialization* is one of these leading trends, encroaching on practically all spheres of human life.

The process of financialization can be described most briefly (and superficially) as the growing importance of the so-called financial services in national economies. The American economist Greta Krippner, in a study of the financialization of the US economy in the post-1970s period, defines the phenomena as a 'pattern of accumulation in which profit making occurs increasingly through financial channels rather than through trade and commodity production' (2005, 181). As she argues, there are two wide-ranging sociological implications of financialization: dubious corporate control and eroded state autonomy. The tendency to global action is another implicit feature of financialization, and as correctly observed by the Dutch political scientist Natascha van der Zwan, 'financialization and globalization are [. . .] not mutually exclusive analytical frameworks, but rather two sides of the same coin' (Zwan 2014, 104). Gradually, the initially useful financial services sector turned into a master of the existing mode of production, leading to its marginalization.

Over the past ten years the work of the Geneva-based Observatoire de la Finance, directed by the Polish-Swiss economist Paul Dembinski, has revealed ubiquitous malignant effects of the expansion of financial logic throughout society, where trust, empathy and responsiveness have been replaced by transactions and contracts: students become 'recipients of educational services', patients 'recipients of medical services', and so forth. Ultimately, the economic field has been twisted and detached from society,[1] thus changing fundamentally the lifeworld as a whole: the sense of public good has been lost, resulting in social unrest – triggered by excessive concentration of wealth (Dembinski 2009, 8) and new global inequalities (Milanovic 2005, Stiglitz 2012, Piketty 2013). Financialization can also be tracked in urban social relations, which for centuries required discussions, negotiations and co-operation but now have been reduced to sale and purchase, while transactions have become the model of increasingly instrumental interpersonal connections. Thus the urban condition does not merely reflect or represent the culture of financial accumulation, but enables the permeation of finance into the fabric of daily life (Moreno 2014).

As noted earlier, financialization conquering the world has been described as *globalization*, one of the most powerful and broadly discussed ideas since the 1980s. By and large, globalization as a concept refers to 'the compression of the world and the intensification of the consciousness of the world as a whole', according to one of the earliest definitions given by the American sociologist Roland Robertson at the beginning of the 1990s (Robertson 2000, 8). Revealingly, the later publications on the topic deal increasingly with what Robertson called 'differentiation' of the globalization, many of them getting more pessimistic and critical of its social and cultural effects and of the lost possible benefits from this course of development.[2] *Westernization, Americanization, Mondialization, glocalization, McDonaldization, McGuggenization, Disneyfication* – these are only few of the numerous sobriquets of *globalization* focusing on the various social, geopolitical, economic and cultural aspects, alternative variations and structural modifications of it. They give us an inkling of the hidden tensions stemming from the new redistribution and upscaling of world power through capitals, information, knowledge, images and lifestyles, but also through new social, cultural and spatial inequalities in the continuously standardized world. As the urban sociologist Sharon Zukin remarked in her already classic work *The Cultures of Cities*, by that time a new symbolic economy had emerged, based on tourism, media and entertainment, and, 'with the disappearance of local manufacturing industries and periodic crises in government and finance, culture is more and more the business of cities' (Zukin 1995, 1). It is also 'a crucial weapon for reasserting order' (ibid., 46). In a later study of the effects of globalization on urban cultures, Zukin suggests that 'Globalization may turn out to be a cruel Darwinian evolution, selecting certain elements of cities [. . .] for preservation while blending others into hybrid, fusion, or "global" forms' (Zukin 2012, 31). Furthermore, challenging global order is also a challenge to conventions regulating the use of spaces: challenging the dominant view of who is permitted to occupy them, for how long, and for which goals, may provoke a brutal response (Zukin, current volume).

An outgrowth of the modernization of societies – driven by the zeal of tycoons for maximizing profit, enabled by mobile technologies and mass communications and benefiting from liberalized markets and democratized border regimes – globalization marked a new era of hyperpowerful multinational economic actors, transcending states and subverting the integrity of nations. This also led to the repositioning of cities and especially large metropolises on the map of global wealth and power. The 'global city', a topic most systematically discussed in the work of the Dutch-American sociologist and political economist Saskia Sassen and interpreted as a node of the international flow of capital, goods, information, labour, migrations and cultures, has turned into the contested terrain of new struggles for space (Sassen 2001, Shaw 2001). 'Furthermore, insofar as powerful global actors are making increasing demands on urban space and thereby displacing less-powerful users, urban space becomes politicized in the process of rebuilding itself' (Sassen 2006, 29).

The politicization of public space is expressed not only in the emergence of an almighty global elite, imposing their visions on city development, management and usage, but also in the appearance of new kinds of global activists on behalf of the powerless and deprived 'new poors', reclaiming not just their rights on the city but their lives as before – their lost jobs and foreclosed homes. One of the most distinct voices of the Occupy Wall Street movement, that of the feminist philosopher and political activist Judith Butler, determined the demands of the protesters in a memorable way: 'If hope is an impossible demand, then we demand the impossible – that the right to shelter, food and employment are impossible demands, then we demand the impossible' (Elliott 2011).

Finally, by the time the processes of globalization became discernible and turned into an object of intensive analyses, the stifling breath of social precariousness and ecological decline was spreading throughout the world. As Roland Robertson said, 'We have entered the phase of what appears to us in 1990 to be the great global uncertainty' (Robertson 2000, 50). However, it was the German sociologist Ulrich Beck who first offered a systematic diagnosis of this shift with the concept of the 'world risk society'. With his book, written as an immediate response to the Chernobyl nuclear reactor catastrophe, he said that modernity entered into such a society when it began to produce globally greater risks than wealth (Beck 1986). While social risks and natural hazards always existed in human history, the late modernity is marked by a fundamentally new situation when the risks are produced as a side-effect of its development: over a long term, the produced risks are global, unpredictable, incalculable and unescapable. At the beginning of the twenty-first century, worldwide uncertainty has become part of the human condition. A decade later, Ulrich Beck added to the already-distinguished environmental and financial risks the terrorist threats, 'which are both empowered and disempowered by the states' (2009, 13). Ultimately, they 'may give rise to a more acute global normative awareness, create a public space and perhaps even a cosmopolitan outlook' (ibid., 15).

The transformed human condition prompts the emergence of what Beck calls '*responsible* modernity' (ibid., 9): a society, globally threatened by natural hazards and political, economic and social risks, gradually developing new ethics and a system of responsibility, substituting reckless consumerism with ever-growing self-restraint, awakening its consciousness and personal responsibility towards nature, and slowly discovering the meaning of a new 'common good', understood as avoiding the possible 'common bad' (Hristova, this volume).

This emerging social reality (with new forms of social organization and new kinds of sociability as well) is denoted differently – as a 'global', 'network', or 'risk' society, but also as a 'resilient', 'post-consumerist', 'post-growth', even 'sustainable' one – each of these terms signifying different aspects of an ongoing worldwide transformation towards a highly reflexive, technologically advanced and increasingly self-organized society that has recognized its own limits, the limits of profit-led production[3] and the limits of its own habitat, but is globally connected and increasingly empowered by the almost limitless possibilities of digital communication systems.

These contradictory trends are projected onto the public space of modern cities – in its further privatization on the one hand, and in the rise of new publics and counter-publics on the other. We can describe them as new kinds of social actors, who are increasingly self-conscious, participatory minded and increasingly proactive, with their new globally valid claims, trans-border mobilization and mundane co-operation. As noted by the American cultural anthropologist Sherry Ortner, there are various truly creative and transformative ways 'in which resistance can be more than opposition' (Ortner 1995, 191).

We observe nowadays a really intensive 'production of space', in the words of Lefebvre (1991), with new representations seeking new ideological closures (Laclau 1990) and new stabilizations (de Certeau 1985). Today we are used to thinking of public space as part of the citizens' rights to the city, 'a contentious site as well as a unifying symbol of civil society' (Zukin, this volume). The global crisis which started in 2008 as a collapse of major financial institutions in the US just demonstrated the fact that the world hegemony reached this point when it brought about global counter-hegemonic responses by a newly born majority of the impoverished middle class. The development of global discontent and resistance is exemplified in the proliferation of transnational social and political networks and informal movements, but also in various largely unnoticeable everyday local practices of disagreement looking for cosmopolitan reimagination and validation (Theodossopoulos and Kirsoglou 2013).

These transfers between the local and global imaginations were enabled by the worldwide connector – the internet – and powerfully enforced with the help of increasingly accessible social media. As described by Manuel Castells, the transforming 'network society' was discovering the highly mediated digital culture of flexibility, adaptability, innovation and decentralized performance (Castells 2005). That was a culture centred less on the hierarchies and more on symmetrical communication. Furthermore, the born-by-the-crisis 'new poors', ex-members of the middle class, turned into 'precariats', and although deprived of their houses and savings, they possessed enough cultural and social capital to mobilize on a mass scale. Their discontent poured out into the public spaces of the modern world, but they also moved fluently between these spaces, transmitting information, knowledge and the practices of urban alternatives.

The reclaimed and regained (even temporarily) public spaces as a product of 'power-filled social relations' (Massey 1999, 21) give evidence of these trends connected to the globalization of protest. It was as early as the end of the last century when Kevin Cox drew attention to the interplay between different local interests in the globalizing environment, and the emergence of broader social networks with porous boundaries through which 'spaces of resistance' occur even in the most totalitarian states (Cox 1998, 3). Now these spaces multiply and 'move' throughout the continents: Tahrir and Taksim squares, Puerta del Sol, Wall Street and Sintagma – they all became landmarks of the radicalization of public space and dissemination of a new spatial approach of contestation throughout the globe. At the beginning of the twenty-first century, the transnational public space was born – with transnational actors, globally disseminated practices and self-organized networks.

Saskia Sassen designates this as the 'Global Street', which employs 'fragments of various national and global territories': a new spatial way of protesting and addressing political issues in a search for social justice 'by those who lack access to the established instruments of power within the frame of national sovereign territory', thus making 'novel territory, and thereby a bit of history', using what was previously considered merely ground. In this specific context, for Sassen 'to occupy is to remake, even if temporarily, territory's embedded and often deeply undemocratic logics of power, and redefine the role of citizens, mostly weakened and fatigued after decades of growing inequality and injustice' (Sassen 2012). The new spatial logic also has different timing: this is a long-lasting process which requires hard work to keep the place safe and people engaged, and it is what makes occupation specific and different from traditional demonstration. Finally, although occupy protests are enabled technically (mobilized, organized and coordinated) through social media – and for this reason they are referred to as 'Facebook revolutions' or 'Twitter revolutions' – they are centred in very strong localities, with very material practices. 'On the one hand these protests were much localized – the protesters occupied very specific real spaces – on the other they had a global megaphone effect' (ibid.).

From such a point of view, the Global Street can be thought in the framework of Charles Tilly's analysis as systemic social movement with all its elements: a campaign, defined as a sustained, organized public effort to make collective claims on target audiences; its own space-based 'repertoire'; and 'WUNC', public representations of worthiness, unity, numbers and commitment on the part of campaigners and their constituencies (Tilly 2004, 53). But unlike those social movements whose activities are reminiscent of 'jazz improvisation' (Tilly 2006, 35), rather than orchestral coordination following previously prepared scenarios, the Global Street and the new transnational public space tend to develop and disseminate their own tactics of protest based on sound analysis and campaigns presented on their own websites. Further on, a number of factors, ranging from technical issues such as openness, accessibility and visibility to the strategic symbolism of the spaces selected as a place for protest, determine the political significance of the occupied urban spaces (Kowalewski 2013).

The trend of the transnationalization of public space has different forms, instruments of expression and intensity. They can vary from social, political and ecological movements to 'new urban wars [. . .] between criminal, paramilitary, and civilian militias, tied in obscure ways to transnational, economic and political forces' (Appadurai 2009, 193) to less dramatic and more routine confrontations as daily strategies and weapons of the weak (Theodossopoulos 2010, 14).

In this vein, this book is constructed around two major narratives – occupation and reimagination – which reflect the most important challenges to public space displayed during the global crisis. The occupation discourse is about the new spatial forms of contesting the established order, including changed forms of ownership, management, uses and reclamation of the space. The other storyline embedded (explicitly and implicitly) in most of the book's chapters is created around reimagination – most generally, the process of rethinking reality in order to reform it. When applied to public space, it pertains to new ways of seeing the place by ascribing to it new functions and uses. Obviously, these two processes, occupation and reimagination, are intertwined and lead to reforming the public space in all its possible aspects: the rules of its access and management, its uses, and its publics with their new 'public good' escaping the universal 'public bad' (Hristova 2014).

As we live in an image-dominated civilization, public space turns into a global screen (to use Lefebvre's metaphor) where images and signs superimpose one over another as instruments of both reinforcement and contest of the existing scopic regimes of surveillance, control and domination. For the Polish philosopher Zbigniew Mikołejko, the world, on one hand, is 'visualized', and on the other hand 'financialized' (Mikołejko 2013). Or vice versa – visualization is one of the instruments for imposing the financial globalism in its numerous manifestations whose final result is the 'homogenized plastic entertainment culture' (Zukin 2012, 12).

In the present book *reimagination* is conceived as both the redesign and remaking of public space via the means of the arts, architecture and civic creativity; and as social imagery about the desired society, imposed in the space through different individual and collective activities, ranging from the official visions of planners, architects and developers to bottom-up civic undertakings, presented in the current volume; for example, those initiated by volunteers, such as the development of the High Line public park in New York (Sharon Zukin) or various urban gardening initiatives in Vienna (Philipp Rode and Eva Schwab).

However, there is a difference between the imagined and the visualized: what is visualized is by necessity pre-imagined, while the imagined is neither necessarily visualized nor always visible. Public space, however, is where these two modes of human creativity make a legitimate couple. As many of the contributors to this volume assert, the changes in the visual public environment imply also a change of its meanings, collective identities and social visions of the citizens (e.g. the chapters of Jerome Krase, Mariusz Czepczyński, Megan Dixon, Wael Salah Fahmi, Philipp Rode and Eva Schwab, among others).

Although the history of imagining, i.e. the individual creative capacity of generating mental pictures as an act of foreseeing and self-organizing, goes back to the early prehistoric societies, only recently has it turned into widely spread social production. Arjun Appadurai was the analyst who first conceptualized the theme of image creation as a collective everyday practice, characterizing the global world of late modernity. He explained the birth of collective imagining as a result of two seemingly quite different phenomena – electronic mediation and mass migration – which both 'impel (and sometimes compel) the work of the imagination' (Appadurai 2009, 4). Following the

Durkheimian tradition, Appadurai defines imagination in the 'postelectronic world' as a collective social fact, which 'has broken out of the special expressive space of art, myth, and ritual, and now has become a part of the quotidian mental work of ordinary people in many societies' (Appadurai 2009, 5), ultimately turning into a fundamental feature of the global culture by the end of the twentieth and the beginning of the twenty-first centuries:

> The image, the imagined, the imaginary – these are all terms that direct us to something critical and new in global cultural processes: the imagination as a social practice. No longer mere fantasy [. . .], no longer simple escape [. . .], no longer elite pastime [. . .], and no longer mere contemplation [. . .], the imagination has become an organized field of social practices, a form of work (in the sense of both labor and culturally organized practice), and a form of negotiation between sites of agency (individuals) and globally defined fields of possibility. This unleashing of the imagination links the play of pastiche (in some settings) to the terror and coercion of states and their competitors. The imagination is now central to all forms of agency, is itself a social fact, and is the key component of the new global order.
>
> (Appadurai 2009, 31)

For Sharon Zukin, in her earlier works the process of reimagination of public space is connected to the domination of redesign strategies during the 1990s that celebrated a 'purposeful vision of urban leisure' whose 'underlying assumption is that of a paying public, a public that values public space as an object of visual consumption' (Zukin 1995, 29). This ultimately led to total 'visible homogenization' (Zukin 2012, 3) as a result of the growing competition between cities striving for alluring authentic landscapes and employing the public imagery of branding to attract financial investments, affluent consumers and tourists, finally producing more of the same – 'the same conspicuous textual allusions and iconic corporate logos [. . .] Homogenization, then, is attendant to strategic visions of urban growth – and shared by elites who have the economic and political power to impose them on urban public spaces' (ibid., 4).

From such a point of view, the occupation practices from the recent decade present an alternative to the homogenized public space, an attempt to re-order (even though temporarily) the world and to reclaim the basic social contract between citizens and governing elites. Ultimately, occupation and reimagination of public space predispose and reinforce each other.

Within this conceptual framework, the book offers a broad spectrum of cases representing different public spaces throughout the world, with their stories, transformation dilemmas and more or less satisfying solutions, starting from new urban cultures of protest to much milder and more cooperative forms of reimagination of urban environments through civic initiatives, community actions, public arts and urban gardening, among other developments. The authors also use different analytical keys for the interpretation of urban space – from sociological, anthropological and social geography approaches to network and visual analysis, often applied interdisciplinarily.

The first part, 'Concepts and discourses: the resilient public space', outlining the broad theoretical framework of the book, is opened by Sharon Zukin's chapter 'Reimagining civil society: conflict and control in the city's public spaces', offering a deep multi-layered analysis of the pervasive changes of public space related to its privatization in a longer temporal perspective. Zukin traces the extension of private governance over public spaces as a historically new form of social control in response to fears of disorder during the 1970s and 1980s in major cities of the world, especially in North America. To a great extent, that was maintained as an adoption and expansion of the Canadian model of the business improvement district (BID), a process designated by Zukin as the 'early privatization' of the 1990s, when 'privatization was institutionalized by the state and seemed to become unstoppable', and the numerous public-private partnerships resulted in 'manicured, programmed, and entrepreneurial public spaces. Ironically, this led to more effective public use, despite continued criticism that the state was giving control to business interests and catering to the wealthy' (Zukin, current volume). The 'advanced privatization' in the 2010s resulted in the appearance of numerous hybrid spaces, privately owned public parks and plazas ('bonus plazas'), which change the rules of using the space but also pose questions to researchers about redefining 'the publicness' of public space.

As the various instructive examples of different public parks presented in the chapter reveal, ownership is not the only criterion determining the public character of urban space. Nor is the critical publicness – in the Habermasian idea of communicative action – which validates only open and free spaces as truly public. A significant aspect of this transformation is the unleashed civic creativity and power of collective reimagination helping to redesign and reuse neglected or abundant places in a city into vibrant public spaces. This often changes the focus of public activities in the new parks from 'politicized' space and encounters with social difference to more elaborate cultural, recreational and environmental programs, free of charge, attracting socially diverse users/consumers and also civic activists and volunteer 'prosumers' of public space, as the example of High Line Park reveals. Simultaneously, as the author suggests, the growing number of hybrid spaces, from High Line to YouTube, is indicative of the increasing dependence of the public sphere on private capital, and cannot conceal the main question about the right to public space which is often translated in practice as the priorities of some users over others. Thus, paradoxically, these two mutually exclusive trends go together: 'From one point of view, public space enacts a social imaginary in which rights are challenged and publics are expanded' (Iveson 2007). 'On the other hand, everyday uses and behaviours create a depoliticized public space that shapes a less contentious civil society' (Zukin, current volume).

Conceptualizing public space as both a physical landscape and a social imaginary, Zukin introduces the idea that it simultaneously operates in different experiential and ideological dimensions, representing different scales of social order, from the immediate experience of the micro-social world to everyday social and macro-social levels to broad meta-social identities, which are conceived as four relatively distinctive analytical scales. This theoretical distinction, based on multi-factor sociological analysis combined with ethnological descriptions of different exemplary American squares, contributes greatly to deeper understanding of the complexities and contradictions of contemporary public space.

The next chapter, 'Public space in a global world: after the spectacle', by the Bulgarian cultural sociologist Svetlana Hristova, analyzes the transnationalization and radicalization of contemporary post-consumerist public space as a result of the changed conditions which determine – in the spirit of the Habermasian theory – its normal functioning as a healthy regulator of social relationships. According to this theory, public space can fulfil its functions as a generator of critical discourse and societal integration only when a balance of mutual control between the state, private sphere and society is established without the superdomination of any of these three. However, with the development of the consumerist society during the twentieth century, public spaces are drained of public meaning and increasingly reduced to locales of consumption and entertainment. On the other hand, the financial globalism created the conditions of impossible possibilities by disabling the states, distorting the markets, and engendering a new class of 'pauperized cosmopolitans', united globally in a 'digital revolution': for a first time in modern history public space became a stage of transnational actors and transnational actions of global discontent. In this process of transnationalization of public spaces, physical and virtual realities merge into 'networks of alternative communication', resulting in their doubling, extension and mutual reinforcement. This process is analyzed through the archetypical case of Tahrir Square, and substantiated by supervening social movements recognized internationally by their referential public playgrounds – the social protests in Athens on Syntagma Square, the Gezi movement and its encampment of Taksim Square in Istanbul, Occupy Wall Street in Zuccotti Park, and *indignados* in Puerta del Sol in Madrid. They became elements of the symbolic geography of global protest which gained transnational visibility through their digital doubling and image circulation in the global symbolic iconosphere. Thusly public space today outgrows the trimming regime of consumerism, set by the neoliberal economy, and becomes a venue for transnational contestations in different forms and by different publics and counter-publics.

This part concludes with the chapter 'Seeing the local in global cities' by the American professor of sociology Jerome Krase, who addresses the problem of reimagination through visualizing spatial practices of contemporary cities and their public terrains. Visualization that takes place locally, and in this sense contributes to the process of glocalization, is addressed here by focusing on ubiquitous aspects of what Jackson called 'vernacular landscapes', such as commercial signs and graffiti that can be seen as local markers of the process of globalization. The images are discussed as they reflect the symbolic competition between more or less recent migrants to claim 'contested terrains', and, as the author states, people change the meaning of spaces and places by changing their appearance.

The second part, 'Contestations and rights: public and civic', features political, social, visual and semiotic transformation of public spaces. Recent alterations of the image, use and significance of urban squares and spaces are often related to changing relationships of power. That change generates tangible and intangible marks in the city landscape and social milieu, which are examined and analyzed in the following chapters. The part opens with Mariusz Czepczyński's chapter 'Civic landscapes of post-socialist cities: urban movements and the recovery of public spaces'. The Polish author studies transformations of post-communist public places to quasi-public and semi-private spaces in Central and East European cities. The 1990s retreat of the civic landscape was visible in practically every city of the region. Since the early 2000s a growing social demand led to some reclaiming and returning of civic functions to public spaces (Czepczyński 2008). Democratic protests, rallies and manifestations have been accompanied by participative movements and cultural and environmental discourses to better employ and utilize civic values in urban landscapes. The revalorization of the civic functions of public spaces embodies the emerging civic society and, in many cities, becomes a sign of a new civic urban landscape.

A more detailed case study of the post-socialist transformation of public space is presented by Craig Young, Duncan Light and Daniela Dumbrăveanu in 'Public space, memory and protest during post-socialist transformation: the emergence of Piaţa Universităţii (University Square), Bucharest, as a space of protest'. The British–Romanian team explores the development of Piaţa Universităţii in the Romanian capital Bucharest, in order to analyse the role that urban public space plays in society and politics in a post-socialist context. The chapter traces how the square developed as a site of protest from its origins in the context of early twentieth-century nation-building, through the communist period and then into the post-communist context. From being a location not explicitly associated with opposition to power, it developed a complex relationship with different forms of protest that existed under communism, and then became more firmly established as a space of protest following the violent events of the 1989 Romanian 'revolution'. These events, and the way in which they are commemorated and remembered, have shaped Piaţa Universităţii as a key site of protest which continues to function throughout the post-socialist period as an important symbolic focus for expressions of resistance to a succession of regimes.

The examination of the transformation of public squares, but in a different context, continues in the next chapter, prepared by the Turkish researcher Nuray Bayraktar. 'Social characteristics of squares as urban spaces: Ulus and Kızılay squares in Ankara' focuses on social and spatial changes of the Turkish capital's major squares. Since the proclamation of the Republic of Turkey and establishment of the capital in Ankara in 1923, the city has been planned and changed rapidly. The main Atatürk Boulevard connects Ulus Square and extends towards Kızılay Square, forming the major urban axis. Recently, due to rapid urbanization, these places have changed, and their uses and users have also changed. Today, Ulus and Kızılay squares are used by diverse groups living in the same city, who gradually lose their place-based relationships.

The chapter by the Colombia-based Francisco Adolfo García Jerez, 'Order and heterotopia in an urban space: the case of a Spanish square', shows how the concepts of heterotopia and order operate in reality, especially in public spaces in Seville, Spain. This case study examines two very different views on how urban life should be activated in public places. On the one hand, civil servants and wider sectors of the population hope for a total renovation of stigmatized neighbourhoods in deteriorated physical condition, with the goal of restoring 'order and security'. While, on the other hand, counter-hegemonic groups reclaim the existence of truly open and public spaces through an inclusive conception of public space. The contraposition of these two views causes social conflicts for the control and use of certain urban spaces.

The part is closed by the chapter by the Egyptian architect and social researcher Wael Salah Fahmi, 'Contested public spaces and the right to the city: the case of Cairo's historic bazaar'. The chapter examines the contested spatiality within Cairo's historic bazaar in relation to various heritage management policies and local interests, revealing the tensions and conflicts between two contrasting approaches of a tourism-based and a community-based rehabilitation. On the one hand, the official visions ('reimaginations') of the historic district as sanitized, gentrified open museum are seeking increased control over public space with restricted accessibility. On the other hand, the study investigates local people's reactions and expectations as they occupy and appropriate their public spaces for various community activities whilst being confronted with compulsory eviction and a lack of residential security. The findings show that official rehabilitation has favoured more technical aspects of the restoration of historic buildings for future tourism, rather than enhancing the area's authentic socio-cultural characteristics and the interests of local inhabitants.

Practices of public space control and ascendancy, as well as changes, uses, occupations and conversions in various scales and in different urban locations are collected in the third part of the book, 'Management and governance: transformation and control'. Time-space relations and various groups of stakeholders build up the social character of changeable, temporal and physical public spaces. This part reflects the conflict between the different needs, aspirations, prospects and limitations of interested parties. The discussion is opened by the chapter 'The meaning of public space in the context of space-time behaviour in the "network city": from socialist to sociable public space'. The Russian–Dutch team of researchers – Anastasia Moiseeva, Remon Rooij and Harry Timmermans – are interpreting recent transformations of public space in post-socialist cities in the Russian Federation in the context of the new social and spatial patterns of the network society and space-time behaviour in the network city. The selective consumption of places by every individual influences and alters the meaning of public space. In this context, the contemporary design of the 'meaning' of public space should concentrate on the connecting interface between different 'enclaves' – 'space as a system of places' – through which different places relate to each other.

The discussion of the transformation of Russian cities is continued in 'The restructuring of urban public space in the Baltic Pearl' by the American social geographer Megan Dixon, who offers interpretation of the observed process using Henri Lefebvre's triad of 'spatial practices, space of representation, and representation of space', revealing multiple aspects of space. The chapter analyses a development project outside St. Petersburg, Russia, for its potential contributions to the city's fund of public spaces. The project, called the Baltic Pearl, is being financed and partly designed by a consortium of firms from Shanghai, China. While actual construction began in 2007, four iterations of design books, the firm's website, and promotional publications allow examination of proposed 'public' spaces; a Lefebvrian analysis is applied both to the discourse and imagery representing the project. In order to extend the analysis, three comparable projects in Shanghai and Beijing are considered. The chapter notes a tendency to design forms of 'public space' around amenities that are regarded as a global 'language' for such spaces. Employing Lefebvrian analysis allowed the author to evaluate the potential for two common desires for public space: facilitating social and/or political debate and creating connections between citizens.

The Austrian practitioners and researchers Philipp Rode and Eva Schwab, in their chapter 'Public green space in Vienna between utopia and political strategy', examine different approaches to public green space, its emancipatory potential and its reinterpretation. The topic is charged with a wide range of meanings and interpretations, and political and social utopias are addressed. They are manifested through the articulation of needs and demands, which have the potential to renew the idea of public space and to come into conflict with existing utilization and management practices and thereby enrich the urban society. This article illustrates that public green space has a significant position in current social processes, which must be given due consideration in further research. A differentiated view on urban green space and its uses deepens the understanding of the dynamics between social movements and administrative institutionalism.

In the next chapter, the Italian sociologist Michela Semprebon brings to the light the issue of the normative construction of a (public) urban space through the use of policy instruments. Built on qualitative research on immigrants' entrepreneurial activities and local conflicts in Italy, the chapter reveals the workings of spatial intervention in a northern Italian city, Verona. Drawing on the political sociology approach to policy instruments, it proposes a detailed analysis of the legislative debate. It shows how the debate was constrained within the technicalities of the actual process, thus frustrating opportunities to value immigrant entrepreneurs' experiential knowledge and to promote innovation. Its main aim is to provide an example of the interweaving dynamics of policy and politics as entrenched in legislative instruments, while contributing to wider discussion on urban politics and democracy.

The last contribution to this part belongs to the Czech sociologist Pavel Pospěch, with his essay 'Negotiating public space in a shopping mall', which interprets the mall as a semi-public city space. The position of malls in terms of public and private is discussed in relation to other city spaces. It is argued that social control leads to the disappearance of otherness from the malls. Three forms of social control are recognized and described: by means of exclusion and surveillance, by architectural means, and by normalization of social conduct. Following this argument, a process of negotiation is described as a fundamental determinant in the social construction of the mall space and its definition.

The conclusion says contemporary public space continues to be what it has always been – simultaneously a venue for the meeting of strangers and an arena of confrontations which, however, has outgrown the boundaries of the national and the imperatives of consumerist culture. Nowadays public space displays the most aggravated conflicts of the globalized world whose meaning is caught between touristification and terrorization; between the need to reassert the rights to the city of urbanites living in constant risk and the necessity to preserve their safety. But public space also provides opportunities for new participatory urban cultures of collective reimagination. Even less radical, mundane activities reveal the new public interests of the risk society in which not only a new 'normal' of global cosmopolitanism (Zukin, current volume) is seen, but also where a new normality for public behaviours and, therefore, a new normativity of urban coexistence, are renegotiated on a daily basis.

Notes

1 We can take as a metaphor of the nature of financialization the observation of Saskia Sassen: 'New York is the leading global market to trade financial instruments on coffee even though it does not grow a single bean' (Sassen 2006, 27).
2 Our aim is not to complete one more exhausting review of the vast body of literature with alternative stances to globalization which appeared by the end of 1990s and the beginning of 2000s. Here we shall mention just a few randomly selected books, coming from different areas of expertise and from different parts of the world: *Globalization: The Human Consequences* by Zygmunt Bauman (1998); *The Cultures of Globalization*, edited by Frederic Jameson and Masao Miyoshi (1999); *Globalization and Its Discontents* by Joseph E. Stiglitz (2002); Manfred Steger's *Globalization: A Very Short Introduction* (2003); Tzvetan Todorov's *The New World Disorder: Reflections of a European* (2003); and Arjun Appadurai's *Modernity at Large* (2009).
3 To put it differently, the society has recognized the limits of the limitless market fundamentalism. 'Even advocates of a global free market are increasingly expressing openly the suspicion that, after the collapse of communism, only one opponent of the free market remains, namely, the unbridled free market which has shrugged off its responsibility for democracy and society and operates exclusively on the maxim of short-term profit maximization' (Beck 2009, 14).

References

Appadurai, A. (2009). *Modernity at Large: Cultural Dimensions of Globalization*. Minneapolis, MN: University of Minnesota Press.

Beck, U. (1992) [1986]. *Risk Society: Towards a New Modernity*. Thousand Oaks, CA: Sage. (Translated from German *Risikogesellschaft: Auf dem Weg in eine andere Moderne*, 1986.)

Beck, U. (2009). Critical Theory of World Risk Society: A Cosmopolitan Vision. *Constellations, An International Journal of Critical and Democratic Theory*, 16(1): 3–22.

Castells, M. and Cardoso, G. (eds) (2005). *The Network Society: From Knowledge to Policy*. Washington, DC: Johns Hopkins Center for Transatlantic Relations.

Certeau, M. de (1985). Practices of Space. In Blonsky, M. (ed), *On Signs*. Baltimore, MD: Johns Hopkins University Press: 122–145.

Cox, K. R. (1998). Spaces of Dependence, Spaces of Engagement and the Politics of Scale, or: Looking for Local Politics. *Political Geography*, 17(1): 1–23.

Czepczyński, M. (2008). *Cultural Landscapes of Post-Socialist Cities. Representation of Powers and Needs*. Aldershot: Ashgate.

Dembinski, P. H. (2009). *Finance: Servant or Deceiver? Financialization at the Crossroads*. London: Palgrave Macmillan.

Elliott, J. (2011). Judith Butler at Occupy Wall Street. Available at https://web.archive.org/web/20140503002548/http:/www.salon.com/2011/10/24/judith_butler_at_occupy_wall_street/ [accessed January 22, 2015].

Hall, S. (2011). The Neo-Liberal Revolution. *Cultural Studies*, 25(6): 705–728.

Hristova, S. (2014). Between Consumerism and Spectacle: Public Space of Central and East-European Cities. In Gura, R. and Styczynska, N. (eds), *Identités et Espaces Publics Européens*. Paris: L'Harmattan.

Iveson, K. (2007). *Publics and the City*. Malden, MA, and Oxford: Blackwell.

Jackson, J. B. (1964). *Discovering the Vernacular Landscape*. New Haven, NJ: Yale University Press.

Jameson, F. and Miyoshi, M. (eds) (1998). *The Cultures of Globalization*. Durham, NC: Duke University Press.

Kowalewski, M. (2013). Miasto jako arena protestów. *Konteksty Społeczne*, 1(1): 18–24.

Krippner, G. (2005). The Financialization of the American Economy. *Socio-Economic Review*, 3(2): 173–208.

Laclau, E. (1990). *New Reflections on the Revolution of Our Time*. London: Verso.

Lefebvre, H. (1991) [1974]. *The Production of Space*. Trans. by D. Nicholson-Smith. Oxford: Blackwell.

Lefebvre, H. (2003) [1970]. *The Urban Revolution*. Trans. by Robert Bononno. Minneapolis, MN: University of Minnesota Press.

Massey, D. (1999). Imagining Globalisation: Power-Geometries of Time-Space. Hettner Lecture 1998, Department of Geography, Heidelberg. University of Heidelberg.

Mikołejko, Z. (2013). *Jak błądzić skutecznie*. Warsaw: Iskry.

Milanovic, B. (2005). *Worlds Apart: Measuring International and Global Inequality*. Princeton, NJ: Princeton University Press.

Moreno, L. (2014). The Urban Process under Financialised Capitalism. *City: Analysis of Urban Trends, Culture, Theory, Policy, Action*, 18(3): 244–268.

Ortner, S. (1995). Resistance and the Problem of Ethnographic Refusal. *Comparative Studies in Society and History*, 37(1): 173–193.

Peck, J. (2012). Austerity Urbanism. *City: Analysis of Urban Trends, Culture, Theory, Policy, Action*, 16(6): 626–655.

Piketty, T. (2013). *Capital in the Twenty-First Century*. Cambridge, MA: Harvard University Press.

Robertson, R. (2000) [1992]. *Globalization: Social Theory and Global Culture*. Thousand Oaks, CA: Sage.

Sassen, S. (2nd ed., 2001). *The Global City: New York, London, Tokyo*. Princeton, NJ: Princeton University Press.

Sassen, S. (2006). Why Cities Matter. In *Cities. Architecture and Society*, exhibition catalogue of the 10th Architecture Biennale of Venice. Available at: www.saskiasassen.com/PDFs/publications/Why-Cities-Matter.pdf *[accessed July 16, 2015]*.

Sassen, S. (2012). The Global Street or the Democracy of the Powerless. Interview by Łukasz Pawłowski in *Kultura Liberalna*, 163 (8/2012). Available at: http://kulturaliberalna.pl/2012/02/20/the-global-street-or-the-democracy-of-the-powerless/ [accessed March 19, 2015].

Shaw, D. (2001). The Post-Industrial City. In Paddison, R. (ed), *Handbook of Urban Studies*. Thousand Oaks, CA: Sage.

Stiglitz, J. (2002). *Globalization and Its Discontents*. New York, NY: W. W. Norton.

Stiglitz, J. (2012). *The Price of Inequality: How Today's Divided Society Endangers Our Future*. New York, NY: W. W. Norton.

Theodossopoulos, D. and Kirtsoglou E. (eds) (2010). *United in Discontent: Local Responses to Cosmopolitanism and Globalization*. New York, NY: Berghahn Books.

Thévenot, L. (2014). Voicing Concern and Difference: From Public Spaces to Common-Places, *European Journal of Cultural and Political Sociology*, 1(1): 7–34.

Tilly, C. (2004). *Social Movements, 1768–2004*. Boulder, CO: Paradigm.

Tilly, C. (2006). *Regimes and Repertoires*. Chicago, IL: University of Chicago Press.

Zukin, S. (1995). *The Cultures of Cities*. Oxford: Blackwell.

Zukin, S. (1998). Politics and Aesthetics of Public Space: The 'American' Model. *Ciutat real, ciutat ideal. Significat i funció a l'espai urbà modern* [*Real city, ideal city. Signification and function in modern space*]. Barcelona: Centre de Cultura Contemporània de Barcelona.

Zukin, S. (2012). Competitive Globalization and Urban Change: The Allure of Cultural Strategies. In Xiangming Chen and Ahmed Kanna (eds) *Rethinking Global Cities: Insights from Secondary Urban Centers*. New York: Routledge, pp. 17–34.

Zwan, N. van der (2014). State of the Art: Making Sense of Financialization. *Socio-Economic Review*, 12, 99–129.

Part 1

Concepts and discourses

The resilient public space

1 Reimagining civil society

Conflict and control in the city's public spaces

Sharon Zukin

Contemporary ideas of public space reflect an idealized view of the squares and marketplaces of ancient times, where strangers of diverse backgrounds are imagined to mingle freely, debate political views, and engage in a lively bazaar of commercial transactions. Those spaces did, in fact, host a wide range of activities that are common to city life and democratic republics. But they limited access to social groups such as women, slaves, and foreigners who did not enjoy full citizenship rights, and their broad plazas, surrounded by temples and assembly halls for the elite, mobilized awe for, and obedience to, both sacred and secular leaders. From the beginning, then, public space has not only involved collective rituals of display, exchange, and passion, but also issues of social conflict and control.

This should not arouse surprise today, when public space is recognized as a contentious site as well as a unifying symbol of civil society. Protestors often mass in the central public spaces of cities around the world to voice demands for change, from Tiananmen Square in Beijing in 1989 to Taksim Square in Istanbul and Maidan Nezalezhnosti in Kiev in 2013. Occupying these spaces, as suggested by the Occupy movement that spread from New York to London, Hong Kong, and many other cities in recent years, asserts a strong claim of the 'excluded' to be equal members of society, a claim that goes beyond formal state institutions to challenge the entrenched power of capital on the one hand, and autocracy on the other.

Though central places have no monopoly as a site of protest demonstrations, they carry a significant moral weight in the history of the nation. Located at a node of media networks, they also command widespread attention. For both reasons, collective actions in these spaces have a strong potential to shape public opinion. Yet central public spaces near the halls of power are fiercely defended by the forces of order: the military and police. Challenging the dominant view of who is permitted to occupy them, for how long, and for which goals, may provoke a brutal response.

Even in normal times, use of public space is subject to rules and regulations. In principle, everyone can enter without paying a fee or coming under suspicion. But in the heightened expectation of protest demonstrations and terrorist attacks, cameras, public police, and privately employed security guards exercise surveillance and control (Graham 2010). Specific groups like homeless people, teenagers, and ethnic minorities are often targeted for intensive policing and prevented from gathering in parks and on the streets (Davis 1990; Mitchell 2003). Increasingly, private associations dominate public-private partnerships to both finance and manage public spaces, and are able to set additional rules for their use. By law, moreover, local and national governments determine what can be displayed and said. These conditions subvert the integrity of public space; they threaten its meaning as 'open' and 'free,' and raise the risk of 'privatization' (Low and Smith 2006).

Yet at the same time, the public in many societies is becoming more socially and culturally diverse. Both transnational and domestic migration bring many 'strangers' into the urban heart of the nation. Initially, hostile efforts often aim to exclude migrants from gathering in public spaces, whether they are informal vendors, day labourers waiting for a job, or just people who look different, or are dressed differently, from the majority (Greenberg and Lewis 2017).

But struggles against discrimination take place on the streets as well as inside legislative chambers and the courts. Mass demonstrations for immigrants' rights use the same symbolically charged public spaces as every other form of political association: in front of city hall, on the steps of the national capitol, in the central public squares. Gradually, just as native-born ethnic minorities are integrated into the everyday life of shopping streets and public parks, partly by changes of law and partly by changes of habit, so transnational migrants also find acceptance. Over time, public parks, shopping streets, and community gardens establish a 'new normal' of cosmopolitan tolerance (Anderson 2011; Amin 2012; Hall 2012).

It is not easy to guess the future evolution of public space, for contradictory pressures will continue to grow to make it more "defensible" but also more accessible and socially inclusive. Among the most important trends of social control, however, are privatization in increasingly entrepreneurial forms and incorporation of difference on multiple scales. These have significant influence on shaping the norms and obligations of civil society.

Privatization of publicly owned space

The extension of private governance over public spaces is a historically new form of social control, which emerged in fears of disorder during the 1970s and 1980s in major cities of the world, especially in North America. During those years, social problems like homelessness and disagreeable, sometimes illegal, behaviour, from urinating and drinking alcohol in public to begging for money and selling proscribed drugs, spread beyond the marginalized urban districts where such 'deviance' had been more or less quarantined.

One factor in the spread of 'deviant' public behaviour was social and political. Increasing poverty, the reduction of the social welfare safety net, and a rising cost of living in the biggest, global cities displaced many people with untreated illness and problems with drug abuse from cheap rented rooms and mental hospitals to the streets and parks. Another factor was cultural. In the aftermath of the 1960s counterculture, public behaviour became looser, more expressive of individual autonomy, and less deferential to authority. Faced with widespread transgression of norms, the police were unable or unwilling to prevent social disorder. Moreover, the financial costs of maintaining order, as well as maintaining the physical infrastructure of public parks and streets, challenged the financial resources of local governments (Zukin 2010).

In New York and other U.S. cities, corporate elites with downtown property interests formed not-for-profit business improvement districts, modelled on a Canadian form of governance, that gave them the right to manage public spaces – mainly shopping streets and public parks – if they paid for their cleaning, policing, landscaping, and promotion. Over time, more local governments were attracted by the opportunity to shift these responsibilities to the private sector. Gradually, the BID model was adopted in cities around the world, a public policy of 'mobile urbanism' that was adapted to local conditions (McCann and Ward 2011).

Beginning with the necessary tasks of keeping public space clean and safe, however, BIDs and public parks conservancies also assumed a 'civilizing mission'. They aimed to make public spaces into havens of consumption by eliminating all potentially unpleasant encounters.

BIDs do this by explicitly banning disturbing behaviour, from homeless people picking glass bottles out of trash cans for a small cash return to skateboarders zooming through pedestrians on the sidewalk. But they also commercialize and democratize an elitist ideal of consumption, a policy that I have called 'pacification by cappuccino' (Zukin 1995). Most importantly, as BIDs have expanded since the 1980s, their upgrading of public spaces has relied on a comprehensive infrastructure of cultural strategies backed by the public, private, and non-profit sectors: street festivals, cultural heritage, farmers' markets, and aesthetically pleasing landscapes. Together, these strategies establish an 'archipelago' of harmonious experiences (Hajer and Reijndorp 2001). While this reshaping of the urban environment can be seen as the result of a concerted ideological campaign against the poor by corporate elites and the more affluent middle class, it is most directly related to the cultural claims of gentrification and the political economy of upscale redevelopment (Smith 1996; Zukin 2010).

The steady advance toward privatized public spaces in the U.S. was marked, at the beginning, by violent struggles over Tompkins Square Park in the East Village of New York City and People's Park in Berkeley, California, around 1990, and, most recently, by the very successful opening of the High Line and Brooklyn Bridge Park, both in New York City, around 2010. During these two decades, privatization was institutionalized by the state and seemed to become unstoppable. From post-industrial abandonment and encampments of squatters, a number of the city's public parks were transformed by public-private partnerships into manicured, programmed, and entrepreneurial public spaces. Ironically, this led to more effective public use, despite continued criticism that the state was giving control to business interests and catering to the wealthy.

Early privatization in the 1990s

For many people in both the East Village and Berkeley, local public parks at the twentieth century's end represented a rare, remaining 'free' space where homeless men and women could camp, self-declared anarchists and political critics could speak, and musicians could play instruments at all hours. Those who advocated this vision were the heirs of the 1960s counterculture and Free Speech Movement. For them, each park 'was working as it should: as truly a public space. . .a political space that encouraged unmediated interaction, a place where the power of the state (and other property owners) could be kept at bay' (Mitchell 2003, 123). When the police tried to exercise control by imposing a curfew, as they did in Tompkins Square Park in 1988, or removing the homeless, as they did in both parks in the early 1990s, protests erupted, although many other community members, especially families with children and local business owners, felt the parks had become too dirty and dangerous.

After 1990, local authorities in the East Village and Berkeley put the two parks on different paths. People's Park languished under the management of the University of California, which issued periodic plans to review the park's use and build facilities for students in it. Faced with a hard core of protestors, however, and an activist organization that fed the homeless there, the university administration allowed the park to remain as it was. This aroused strong emotions on all sides. On the one hand, supporters of the 'free' space cheered it as an example of social inclusiveness and praised it for giving more privileged people an opportunity to experience poverty and social difference. On the other hand, critics avoided going into the park, fearing its homeless residents' dogs, illegal drugs, and aggressive verbal and physical attacks.

In New York, the liberal city government insisted on removing the homeless encampment from Tompkins Square Park. Then the parks department closed it for a year to carry out an extensive new design and renovation. 'This park is a park,' Mayor David N. Dinkins said. 'It is not a place to live. I will not have it any other way.' The parks commissioner shared this view. 'What we've learned is when you start to see that stuff – people putting up tents and tepees – you've got to go in and get rid of it,' Commissioner Betsy Gotbaum said. 'Otherwise it will turn into a shantytown, and that's not what parks are for' (Kifner 1991).

Some East Village community members endorsed this approach, forming an association to build a dog run in the park, and later raising nearly half a million dollars to improve it. Local volunteer work and fund raising to build and maintain facilities were very much in line with the citywide movement to establish BIDs and public parks conservancies. The combination of user-volunteers and public-private partnerships was championed by Dinkins' successor as mayor, Rudolph Giuliani (1994–2001), and greatly expanded under his successor, Michael Bloomberg (2002–13).[1]

In 1995, three local restaurant owners formed the East Village Parks Conservancy, which took as its mission, in partnership with the parks department, 'to clean Tompkins Square and add plants' (www.evpcnyc.org/index2.html, accessed August 1, 2014). Today, stewardship of public parkland remains the conservancy's goal, aided by fund raising and organizing volunteers to keep the park clean. The conservancy even sells sponsorships of individual trees, including, for the biggest donations, the park's three historic Great Sycamores. Unlike People's Park, Tompkins Square Park hosts free summer movies and an annual jazz festival, as well as offering basketball courts, children's playgrounds, a weekend farmers' market, and lush green lawns. In contrast to Berkeley residents' continued avoidance of People's Park, East Village residents make daily use of Tompkins Square Park.

To some degree, generational differences separate the old counterculture from today's park users. But it cannot be denied that since the 1990s, both Berkeley and the East Village have experienced gentrification. In general, a majority of community residents in both places, including large numbers of university students, seem to prefer a well-managed park to a place that forces them to confront social problems. Yet a refurbished public park helps to raise rents and attract upscale cafés and shops, and these changes may not benefit long-time residents or students. Even more problematic, the governance model of BIDs and parks conservancies reduces the local state's responsibility for maintenance. As a result, public space must depend on generating for-profit activities.

The dramatic success of two new public parks – the High Line, a 'vertical park' that was created on an unused, elevated rail freight line overlooking the Hudson River, and the Brooklyn Bridge Park, built on the site of abandoned piers in the East River – shows how influential this entrepreneurial model has become.

Advanced privatization in the 2010s

The High Line resulted from the civic activism of volunteers who imagined that an elevated walkway near the river could attract visitors by adopting traditional forms of passive urban recreation: strolling, people watching, and gazing at the view. At first, during the 1990s, the site's lonely abandonment only attracted the interest of some artists and historic preservationists. But by 2002, handsome photographs of wild grasses growing through cracks in the cement excited a broader fascination. The photographs became a means of promoting the rail line's reuse, especially since several hundred art galleries and a number of trendy restaurants and boutiques had already moved into the adjacent areas of Chelsea and the Meatpacking District (www.youtube.com/watch?v=lNzr7g8FQgk, video made 2002, accessed August 4, 2014; Sternfeld 2012).

The proposal to create a new kind of park appealed to creative-sector entrepreneurs who worked and lived in this part of the city, and also complemented the plans of real-estate developers who saw investment potential along the West Side (Halle and Tiso 2014). With their financial donations and political support, a volunteer association of activists, Friends of the High Line, campaigned to transform the viaduct into a public park. They organized a high-profile design competition to attract attention, spoke tirelessly about the economic value that a new public park would create, and finally, when the media and real-estate development began to kick in, and the Bloomberg administration was eyeing the West Side as the site of a future Olympics stadium, they persuaded the mayor and city council to join them.

In 2009, ten years after the Friends began their campaign, the High Line opened as a public park. Though initially connected with the city's failed bid to host the 2012 Olympics, the park soon become a major tourist attraction in its own right. Media from around the world praised its simple yet elegant landscaping, expansive views of the river, and places to sit and observe the passing crowd. As a public space, the High Line was a *flâneur*'s dream.

Yet the park also anchored the massive development of luxury apartments and high-status offices. Offices immediately adjacent to it commanded rents that were 51 percent higher than in similar buildings one block away (Kusisto 2012). So it wasn't just love of green space that motivated huge financial donations to the High Line Improvement Fund from real-estate developers, nearby corporations, and individual philanthropists. Developers even got a 'bonus' from making a donation. In the Special West Chelsea district that was established in 2005, developers who donated at least $50 per square foot of their planned floor area were rewarded with permission to build more rentable space than normal zoning laws would allow (New York City 2005). This caused money to flow to the High Line, as much as $85 million in 2011, placing it among a small number of 'elite' public parks in the city (Lindner and Rosa 2017).

At the same time, when Brooklyn Bridge Park was created in 2002, it applied the entrepreneurial model to a wider range of for-profit activities. In a striking innovation, New York City and New York State required this park to be financially self-sustaining.

Like the High Line, Brooklyn Bridge Park is located on a waterfront. It also began as a community activists' vision of how old industrial-era infrastructure could be refashioned and reused. But in this case, unused piers would be transformed into parkland instead of being demolished, and part of the site would be used for intensive residential and commercial real-estate development. A public park would not only offer green space for recreation, it would protect local residents' fabulous views of the East River, the Statue of Liberty, and the Lower Manhattan skyline (Hand and Pratt Pearsall 2014).

During the 1980s and 1990s, community organizations in Brooklyn Heights, the affluent neighbourhood adjacent to the waterfront, pressed elected officials to support this vision. In principle, New York City and State liked green space, but the Port Authority of New York and New Jersey, which managed the piers, wanted to raise revenues by leasing the site to real-estate developers. Pressure to capitalize on the space was intense, for the city administration was still trying to overcome a severe fiscal crisis. Moreover, from 1998 the city, state, and federal government were contributing funds to the long-term development of a much larger park along the Hudson River.

For all these reasons, public officials sought creative financing that would make Brooklyn Bridge Park monetarily self-sustaining. The strategy that they devised put the non-profit management of the park directly in charge of real-estate development (www.brooklynbridgepark.org, accessed August 2, 2014).

New York State gave the park developable 'parcels' of land and buildings both on and adjacent to the waterfront, and charged the Brooklyn Bridge Park Development Corporation with redeveloping them. BBPDC lays out the parameters of each development project, solicits and evaluates proposals from real-estate developers, and receives revenues from leasing and PILOT fees (payment-in-lieu-of-taxes). Presumably, the development corporation selects the highest bids and balances projected revenues with public benefits. As a result, the park is gradually being surrounded by apartment houses, a shopping centre and offices, a theatre, and a hotel, which pay for the 85-acre park's 'world-class' landscaping and recreational facilities. Despite these extremely attractive amenities, the specific arrangements for development have caused endless conflict.

Aside from the environmental impact of building on the waterfront in the face of climate change, some people question whether housing should be built in a public park at all. Then there is the question: building housing for whom? During the Bloomberg administration, all the apartments planned for the waterfront were to be luxury housing, in order to maximize the park's revenue. But when the populist de Blasio administration took office in 2014, plans changed. Now, 30 percent of the apartments in the residential developments are required to rent at 'affordable' price levels; the other 70 percent will rent at market rates. Residents of the first apartment house developed on the park, along with homeowners in Brooklyn Heights, are seeking to overturn these plans. Some don't want lower-income residents to move into apartments near them, while others object that the proposed new towers will block their view. Opponents also argue that publicly subsidized rents will not contribute to the park's revenues (Robbins 2014).

The High Line and Brooklyn Bridge Park have moved a long way from the issues that roiled community residents in Berkeley and the East Village in the 1990s. Instead of 'politicized' space and encounters with social difference, the new parks offer elaborate cultural, recreational, and environmental programs, free of charge, to a socially diverse array of users. And instead of challenging the right of homeless men and women to build a camp in a public park, some community residents now argue against building apartments for lower-income neighbours.

Each park relies on professional, non-profit management from outside the public sector and an army of voluntary donors and 'friends'. But the underlying issue remains privilege. Should donors who fund public space reap financial benefits? Who has the right to use it? Do some users have priority over others?

It is true that both parks are tremendously popular. More than four million visitors walk on the High Line each year; more than 100,000 visitors come to Brooklyn Bridge Park on a summer weekend. Yet in both cases, privatization of public space has become the governance model of choice. It is institutionalized in the public-private partnerships that manage the parks. It takes monetary form in the donations that the parks solicit and the revenues that Brooklyn Bridge Park gains from development. Parkland, which in many ways is the least pecuniary form of public space, has been made to conform to the increasingly entrepreneurial orientation of local government (Harvey 1989).

In principle, at least public parks are still publicly owned.[2] But growing numbers of hybrid spaces, from the High Line to YouTube, hide the increasing dependence of the public sphere on private capital.

Privately owned public plazas

Zuccotti Park, where the Occupy Wall Street protesters camped for several months in 2011, is one of these hybrid spaces. It looks like a public park and is required, by law, to be open to public use 24 hours a day. But in fact it is a privately owned public space, one of more than five hundred in New York City (Foderaro 2011).

Built and maintained by developers, privately owned public plazas were created by New York City's Zoning Resolution of 1961 as a trade-off between public benefits and private profits. In return for including a public plaza at ground level, either indoors or outside, developers get a 'bonus': they are allowed to build more rentable floor area than zoning laws permit in that location. For the most part, this trade-off results in taller buildings in already densely built commercial districts such as midtown and Lower Manhattan. But the plazas do provide breathing space, with sunlight and air. For this reason, they are considered to make a positive contribution to public health and urban design (www1.nyc.gov/site/planning/plans/pops/pops.page, accessed May 20, 2017).

Because developers retain the right to shape their public plazas, they use them to enhance the financial value of their buildings. In principle, this public space is open to everyone during all hours that the building is open, and a sign is posted to make that access clear. In practice, however, developers often instruct their architects to make the public space uncomfortable or even unattractive, in order to discourage homeless, unruly, or merely dishevelled visitors from lingering there (Smithsimon 2008; Németh 2009).

In 2007, following an authoritative report on New York's privately owned public spaces that documented, in many cases, their forbidding look and utter uselessness (Kayden 2000), the city planning commission adopted explicit new rules for their design. They first aimed to eliminate 'sunken' and elevated plazas which were built either below or above street level, and therefore looked both unattractive and inaccessible. 'A public plaza must be visually interesting and easily seen from the street,' the new rules begin, giving 'evidence that it is an open, public space' (www1.nyc.gov/site/planning/zoning/districts-tools/private-owned-public-spaces.page, accessed May 20, 2017). The plaza must also include comfortable seating, and be 'visibly connected to the street' in order to provide users with a sense of safety and security.

As a result of the new guidelines, many bonus plazas are better designed. But developers' willingness to make them open and attractive may reflect lower crime rates citywide in recent years, as well as greater reliance on private security guards, more extensive electronic surveillance, and aggressive removal of homeless people. Moreover, researchers note, new plazas have more physical barriers, like gates, and post more rules and regulations about appropriate behaviour than older ones (Schmidt, Németh and Botsford 2011).

Yet the more 'open' a privately owned public space appears, the greater the perceived need for transparency. In 2011, when Occupy Wall Street protestors camped in Zuccotti Park, the development firm that owns the space hesitated for two months to ask the police to remove them, from concern about the legality of such action and perhaps also from anxiety about their public image. Nonetheless, a New York criminal court judge ruled in 2012 that a developer has a legal right to remove users of the public space that they own, if conditions become dangerous. In Zuccotti Park, the fire department argued that protestors and their tents blocked the park's exits, posing a danger to human life in case of fire (Moynihan 2012).

Hesitation to move against the Occupiers, at first, did not mean that the city government had developed a more tolerant understanding of the right to protest in public space. Indeed, since an extensive security zone was established in Lower Manhattan following the terrorist attack on the World Trade Center on September 11, 2001, and protestors were arrested during the Republican National Convention in midtown Manhattan in August, 2004, the police department has not hesitated to limit protests in public space by all available means. First, the police department tries to pre-empt protests by denying applications for permits to assemble and march in the streets and public parks. Then, during political demonstrations, police officers physically harass protestors, confine them to narrow passages between steel barricades, and make large numbers of arrests – even if demonstrators are not breaking any laws (www.nyclu.org/RNCdocs, accessed August 5, 2014).

As all of these examples suggest, the social production of public space responds to a complicated matrix of general forces like power and money; specific pressures, including protests and litigation; and trade-offs between amenities and security. Like any social space, public space is not a passive location; it is an active agent in producing civil society. But the processes that public space enables and enfolds have contradictory effects. From one point of view, public space enacts a social imaginary in which rights are challenged and publics are expanded (Iveson 2007). On the other hand, everyday uses and behaviours create a depoliticized public space that shapes a less contentious civil society.

The question of social order

During the 1980s, when new public-private partnerships rewrote the rules of public space, they relied on a dramatically different idea of how social order is maintained that was described by the influential New York writer William H. Whyte. Instead of focusing on either the coercive power of the police or the 'defensive' power of design, Whyte (1980) called attention to the way park users behave when left to themselves, creating in a positive sense 'the social life of small, urban spaces.' This idea was based on systematic observation of users in a variety of public parks in several U.S. cities, as well as in Paris.

Whyte and his team of researchers discovered that park users who behave well, who uphold the dominant norms of a civilized, social order, set the standard for others. More surprisingly, they saw that the more pleasant and permissive the environment, the greater the users' sense of ease and the lower the risk of physical attacks, theft, or vandalism. In a major departure from common practice at the time, Whyte recommended using movable tables and chairs rather than bolting benches to the ground.

Whyte also found that specific design features enhance security. Most importantly, he suggested removing tall trees and bushes in order to make everyone in the park visible to each other as well as to passers-by in the street. Clear sightlines would enable mutual surveillance. Whyte proposed, in other words, an urban panopticon, to achieve the greatest public benefit with the least overt repression.

It was a remarkably confident vision, especially at a time when so many people were anxious about links between disorder and violent crime. But it also coincided with the emergence of the 'broken windows' theory of policing (Kelling and Wilson 1982), which emphasized the need to preserve social order by both keeping physical facilities in good repair and catching people who commit small crimes before they 'move on' to bigger ones. Unlike Whyte's ideas, the broken windows theory led not to more pleasant public spaces, but to more pre-emptive policing.

Whyte's recommendations were adopted by civic associations like the Project for Public Spaces and the Municipal Art Society that aimed to improve the quality of urban life in a voluntaristic way outside state control. On this point, Whyte followed the popular urban writer Jane Jacobs's (1961) views. Jacobs had written about store owners and residents who informally watch over a neighbourhood by looking out their window and checking that everything is all right. Their 'eyes on the street', she argued, are the most effective deterrent to crime. Twenty years later, these ideas were codified by the city planning commission's design rules for bonus plazas and put into practice in many public parks. As subtle strategies of social control, they remain widely influential today.

But 'the social life of small urban spaces' operates only in the here and now of co-present sociality. In contrast, as I have shown so far, it is useful to think of public space as operating according to different kinds of rules, on physical access, activities, and ownership (cf. Németh 2012). Yet if public space is not only a physical landscape, but also a social imaginary, it is important to understand that it simultaneously operates in different experiential and ideological dimensions, and on several different analytic scales.

Scales of social order

Scales of social order are not necessarily consistent within a single public space. Subjective values of users may disrupt the objective rules that are posted by management. The degree of 'order' may also vary across different kinds of public spaces, from the relative chaos of a city street to the enforced coherence of a privately owned bonus plaza. To simplify, let's see how public space operates on four scales of social order, from the immediate experience of the micro-social world to broad meta-social identities.

Beginning with the *micro-social level*, the authorities that manage a public space have direct control over users' behaviour. The coercive power of the state, operationalized by the police, offers the clearest and most common examples, such as denial of a use permit to hold a protest demonstration, removal of 'dangerous' users, and outright bans on dangerous or annoying behaviour like skateboarding or panhandling. Slightly more subtle are posted rules of use, like the signs outside bonus plazas that state the hours when the space is open to the public, and the comprehensive signs outside public parks that prohibit smoking, littering, panhandling, walking dogs without a leash, using illegal drugs or alcohol, entering after the park is closed for the night, poking through the trash cans, and 'performing or rallying, except by permit.'

Certainly, rules may be enforced selectively, discriminating against some social groups while favouring others. In Manhattan's Bryant Park, despite posted rules prohibiting the consumption of alcohol, the BID's private security guards ignore young office workers and professionals who drink cocktails on summer evenings while picnicking on the grass (Buckley 2008). However, men and women who drink alcohol from bottles hidden in brown paper bags – an illegal act in many North American cities, and one associated with a lower social class – can expect to be evicted from the park or even arrested.

Yet the micro-social order of a public space may be constructed to protect vulnerable groups. Beginning in the 1960s, playgrounds in New York City public parks were fenced in by iron gates, and posted signs warned that only children and their caregivers could enter. In some parks, complaints about older children's active style of play led to separate playgrounds for babies and toddlers, on one side of the park, and older children on the other. These practices were based on the architect Oscar Newman's (1972) theory of 'defensible space', combined with the amenities endorsed by Whyte.

Other design features defend public space more aggressively. Since the 1980s and the expansion of the homeless population, street furniture is modified by installing sharp prongs or metal dividers to prevent men and women from sleeping there. Yet some features of the micro-level of social order are not consistent. New York's 'street furniture' includes, on the one hand, backless benches divided by short arm rests which allow passers-by to rest, but not stretch out and go to sleep, and, on the other hand, movable tables and chairs where passers-by can lounge on 'traffic-calming' islands.

On the everyday *social level*, a public space translates institutionalized power relations into physical landscape. This often works by the symbolic appropriation of public space, as when statues of historical figures commemorate formative experiences of the city or nation. These symbols repress memories of some eras while selectively reinforcing others. By contrast, removing statues that have dominated public spaces for years makes a change of regime visible. It also actively invites participation in a new form of civil society (Czepczyński 2008).

In an economic as well as an ideological dimension, a public space confirms social privilege. The views offered by the High Line and Brooklyn Bridge Park are spectacular examples of local privilege, but both parks amplify this advantage by raising property values. The value of public space in the scarcity of nature cannot be challenged. Even small community gardens, which are created and maintained by residents' own labour, raise property values in low-income areas of the city, increasing the prospect of gentrification (Voicu and Been 2008).

Access to public space may reinforce many other kinds of inequality. When increased demand for use of the playing fields in New York's public parks led to conflicts of interest between soccer and baseball teams, between girls' and boys' teams, and between teams from public and private schools, the responses of different social groups reflected their unequal social positions. Native-born Americans who mostly play baseball supported the interests of baseball teams, while immigrants who play soccer demanded equal facilities. In one case, Honduran immigrants in the Bronx pressed the parks department to create a soccer field. But when the field opened, the parks department gave preferred playing time to several youth leagues instead of to the adult, Honduran league (Wall 2012). On a social level, access to public space raises questions about whose interests get priority.

On a *macro-social level*, public spaces are representations of general or collective cultural values; they enable citizens to perform routines of social and environmental sustainability. Public parks, for example, both embody and 'emplace' the green space of metropolitan nature that is always scarce in cities. Parks enable city dwellers to enjoy and support the grass, plants, and trees that refresh the senses and help to sustain the natural environment. In a different way, a farmers' market evokes the market-place in European towns and the Central American plaza, with their cultural values of sociability. A public market enables forms of social interaction that sustain an almost mythical image of community (Low 2000; de la Pradelle 2006).

Yet some concepts of public space contradict others. Is a public park a space of contemplation and repose, or can it also be a market? Do commercial concessions like food stands provide a useful amenity, or do they only tranquilize people into being passive consumers? To some degree, different uses reflect diverse origins of public space, beginning with the dichotomy between the political forum of ancient Rome and the commercial agora of ancient Athens (Weintraub 1994). But they also indicate an incompatibility of means and ends, reflecting tension between aesthetic and ethical values.

Finally, on the *meta-social level*, public spaces construct broad social identities. These refer to people's ability to think of themselves as city dwellers, citizens of a nation-state, and free human beings. Being in a public space together with others can bring relief from feeling alone, intense feelings of solidarity, and catharsis after a tragedy. Public space enables men and women to imagine the public sphere as a material place of co-presence, interaction, and transcendence.

But the specific emotional effects of being in public space pull people in different directions. The sense of the crowd on a city street can feel either crushing or dynamic. Some women feel violated by the gaze of men in a public park, while others feel liberated by the opportunity to sit alone on a park bench among strangers. Taking a breath of fresh air in a public park reawakens a human bond with trees and squirrels, while nearby, in another part of the park, a cappuccino taken in a café rekindles the urbane appetite of a *flâneur*. Surely the democratic potential of public space is enriched by these contradictions.

Final observations

For most of us who write about public space, analysis occupies a middle ground between ideology and ethnography. We know what we want public space to be, and we also know what it should feel like.

For me, writing about public space means that I must confront the contradictions of Union Square, the most prominent public space in my neighbourhood. When I walk through it, as I do nearly every day, I cannot avoid brooding on the fact that it is a well-used and well-loved public space, but it is managed by a business association.

Union Square is, in fact, where the first BID in New York State was formed, in the early 1980s, at a time when drug dealers commandeered the small park at the centre of the square, and a sense of disorder hung over the city like a curse. The Union Square Partnership adopted William H. Whyte's optimistic ideas about the social order of small, urban spaces, but they also hired a team of private security guards to work closely with nearby police precincts. From the beginning, there was a trade-off between the amenity of the park and its security.

The Partnership changed the design of the park, removing the high stone walls and tall trees that had made it impossible to see into the park from outside. They installed new flowering plants, and funded long-deferred maintenance. In 1998, the U.S. Parks Service named the renovated park a national historic landmark, in tribute to its history as a gathering place for labour demonstrations. But more New Yorkers appreciate the park because, four days a week, it hosts the city's largest Greenmarket, a farmers' market that features only locally grown food products.

In the park, the BID sponsors free fitness, yoga, and arts and crafts classes. Every Thursday evening in the summer, a different band plays free music and there is dancing. During the day, vendors sell art work, T-shirts, and used CDs on the sidewalk, although since 2010 the parks department has limited the number of vendors to reduce congestion. Often someone plays the bagpipes, a violin, or a piano, hoping to collect money from passers-by. On one recent evening I passed a drummer beating a steady rhythm on an *atabaque* while men, women, and children performed the graceful exercises of capoeira, Brazilian self-defence.

Mostly men play timed chess for money at tables they set up on the 14th Street side of the park every morning. At lunchtime, the park is crowded with people eating lunch on benches and at movable tables and chairs. Every evening, as many as two hundred young people gather on the park's broad front steps to hang out.

I cannot pass Union Square without remembering how people meet there in times of crisis. In the weeks following the attack on the World Trade Center in 2001, New Yorkers gathered together in the park to exorcise their fear. During the five days of a power blackout in Lower Manhattan that followed Hurricane Sandy in 2012, emergency food and water were given out in Union Square Park, and a free charging station was set up there for cell phones and computers.

But all is not calm in paradise. During the past few years, the Partnership was sued by a community group for renting the only roofed structure in the park, the historic stone pavilion, to a sit-down restaurant. Though the restaurant is informal and its prices are reasonable, the group argued that this use of the space excludes members of the public who cannot pay, and therefore 'alienates' publicly owned parkland.

As the name implies, a BID primarily aims to make conditions better for business and commercial building owners. Benefits to the community, or to the wider public, come at the BID's discretion. Like the board of directors of all other BIDs in New York City, the board of the Union Square Partnership is dominated by real-estate developers and agents. As with the High Line and other parks, there is a strong synergy between Union Square's pleasant amenities and property values.

The Union Square Partnership is the Growth Machine of urban economic development in action. Yet the park is one of the most effective public spaces in the city.

Gatherings in public space may remind us of ancient republics, but the power over place and the emotional bonds that are reawakened there are both primal and new.

With so much at stake, conflicts over the management of public space will be ever more important in shaping civil society.

Notes

1 Yet Mayor Giuliani had a political battle with the most prominent BIDs, located in midtown Manhattan. The mayor was angry because the director of three BIDs, who simultaneously held all three executive positions, took strategic actions, such as planning to issue municipal bonds to fund improvements, that threatened the mayor's power – and the city government's. Under pressure by the mayor, the city council made a new law that prohibited holding the directorship of more than one BID, and also brought BIDs under the supervision of the municipal Office of Business Services. For their part, advocacy organizations such as the Coalition for the Homeless and the New York Civil Liberties Union were angry because one of the midtown Manhattan BIDs forced homeless men to take low-paying jobs with the BID as street cleaners and guards. In Archie v. Grand Central Partnership (1995), a judge ruled that the BID had violated federal and state minimum wage rules; the case was settled in 2000 when the BID paid more than $800,000 in back wages.
2 Parkland can be rented by the day for both public and private events. However, state law prohibits the permanent "alienation" of parkland from public ownership, usually by a sale, without specific approval by the New York State Legislature.

References

Amin, A. (2012). *Land of Strangers*. Cambridge: Polity Press.
Anderson, E. (2011). *The Cosmopolitan Canopy: Race and Civility in Everyday Life*. New York, NY: Norton.
Buckley, C. (2008). Ah, the Heat, the Crowd, the Park, and the Booze. *New York Times*, July 16.
Czepczyński, M. (2008). *Cultural Landscapes of Post-Socialist Cities: Representation of Powers and Needs*. Aldershot: Ashgate.
Davis, M. (1990). *City of Quartz: Excavating the Future in Los Angeles*. London: Verso.
Foderaro, L. W. (2011) Privately Owned Park, Open to the Public, May Make Its Own Rules. *New York Times*, October 13.
Graham, S. (2010). *Cities Under Siege: The New Military Urbanism*. London: Verso.
Greenberg, M. and Lewis, P. (eds) (2017). *The City Is the Factory*. Ithaca, NY: Cornell University Press.
Hajer, M. and Reijndorp, A. (2001). *In Search of the New Public Domain*. Rotterdam: NAI.
Hall, S. (2012). *City, Street and Citizen: The Measure of the Ordinary*. London: Routledge.
Halle, D. and Tiso, E. (2014). *New York's New Edge: Contemporary Art, the High Line, and Urban Megaprojects on the Far West Side*. Chicago, IL: University of Chicago Press.
Hand, S. M. and Pratt Pearsall, O. (2014). *The Origins of Brooklyn Bridge Park, 1986–1988*. Memoir on the website of the Brooklyn Historical Society, Brooklyn, NY, http://brooklyn history.org/docs/OriginsBrooklynBridgePark.pdf [accessed August 4, 2014].
Harvey, D. (1989). From Managerialism to Entrepreneurialism: The Transformation in Urban Governance in Late Capitalism. In Harvey, D. *Spaces of Capital*. New York, NY: Routledge, 345–68.
Iveson, K. (2007). *Publics and the City*. Malden, MA, and Oxford: Blackwell.
Jacobs, J. (1961). *The Death and Life of Great American Cities*. New York, NY: Random House.
Kayden, J. S., New York City Department of City Planning, Municipal Arts Society (2000). *Privately Owned Public Space: The New York City Experience*. New York, NY: John Wiley & Sons.

Kelling, G. L. and Wilson, J. Q. (1982). Broken Windows: The Police and Neighborhood Safety. *The Atlantic*, March. www.theatlantic.com/magazine/archive/1982/03/broken-windows/304465/ [accessed August 5, 2014].

Kifner, J. (1991). New York Closes Park to Homeless. *New York Times*, June 4.

Kusisto, L. (2012). Parks Elevate Office Rents. *Wall Street Journal*, August 28.

Lindner, C. and Rosa, B. (eds) (2017). *Deconstructing the High Line: Postindustrial Urbanism and the Rise of the Elevated Park*. New Brunswick, NJ: Rutgers University Press.

Low, S. (2000). *On the Plaza: The Politics of Public Space and Culture*. Austin, TX: University of Texas Press.

Low, S. and Smith, N. (eds) (2006). *The Politics of Public Space*. New York, NY: Routledge.

McCann, E. and Ward, K. (eds) (2011). *Mobile Urbanism: Cities and Policymaking in the Global Age*. Minneapolis, MN: University of Minnesota Press.

Mitchell, D. (2003). *The Right to the City: Social Justice and the Fight for Public Space*. New York, NY: Guilford.

Moynihan, C. (2012). Owner Had Right to Clear Zuccotti Park, Judge Says. *New York Times*, April 8.

Németh, J. (2009). Defining a Public: The Management of Privately Owned Public Space. *Urban Studies*, 46(11): 2463–90.

Németh, J. (2012). Controlling the Commons: How Public Is Public Space? *Urban Affairs Review*, 48(6): 814–838.

Newman, O. (1972). *Defensible Space: Crime Prevention through Urban Design*. New York, NY: Macmillan.

New York City Planning Commission and Department of City Planning (2005). *Zoning Resolution, Web Version, Article IX: Special Purpose Districts, Chapter 8: Special West Chelsea District*, www1.nyc.gov/assets/planning/download/pdf/zoning/zoning-text/art09c 08.pdf [accessed May 11, 2017].

Pradelle, M. de la (2006). *Market Day in Provence*. Trans. by A. Jacobs. Chicago, IL: University of Chicago Press.

Robbins, L. (2014). The Battle of Brooklyn Bridge Park. *New York Times*, August 3.

Schmidt, S., Nemeth, J. and Botsford, E. (2011). The Evolution of Privately Owned Public Spaces in New York City. *Urban Design International*, 16(4): 270–84.

Smith, N. (1996). *The New Urban Frontier: Gentrification and the Revanchist City*. London: Routledge.

Smithsimon, G. (2008). Dispersing the Crowd: Bonus Plazas and the Creation of Public Space. *Urban Affairs Review*, 43(3): 325–50.

Sternfeld, J. (3rd ed. 2012). *Walking the High Line*. Göttingen: Steidl.

Voicu, I. and. Been, V. (2008). The Effect of Community Gardens on Neighboring Property Values. *Real Estate Economics*, 36(2): 241–83.

Wall, P. (2012). Honduran Soccer League Can't Use New Crotona Park Field They Fought For. www.dnainfo.com [accessed August 5, 2014].

Weintraub, J. (1994). Varieties and Vicissitudes of Public Space. In Philip Kasinitz (ed), *Metropolis: Center and Symbol of our Times*. New York, NY: New York University Press, 80–319.

Whyte, W. H. (1980). *The Social Life of Small Urban Spaces*. New York, NY: Project for Public Spaces.

Zukin, S. (1995). *The Cultures of Cities*. Malden, MA, and Oxford: Blackwell.

Zukin, S. (2010). *Naked City: The Death and Life of Authentic Urban Places*. New York, NY: Oxford University Press.

2 Public space in a global world

After the spectacle

Svetlana Hristova

Introduction: public space on the volcano of civilization

We live in a world that is almost cracking under the load of global risks. Today's crisis can be described as a structural effect of the risk society, producing more threats than benefits (Beck 1992): it is multidimensional, encompassing all sectors of social life, and institutional at the same time, affecting major institutions of late-modern culture: cities, nation-states, regions. Consequently, new counter-movements are mobilized, and as Nancy Fraser points out, their 'struggles over nature, social reproduction and global finance constitute the central nodes and flash-points of crisis. On its face, then, today's crisis is plausibly viewed as a second great transformation, a great transformation redux' (Fraser 2014, 544). Most often this situation is interpreted as a consequence of the politics of neoliberalism; as an effect of globalization, and, more recently, as a result of financialization[1] (Dembinski 2009; French et al. 2011, Moreno 2014), revealing the process by which financial logic has subjugated all spheres of social life and colonized the human lifeworld, mutating the DNA of contemporary culture. However, to understand the operating mechanism which systemically engenders crises and produces structural unemployment, we need to consider these twinning concepts in their interrelatedness – as neoliberal financial globalism.

Indisputably, this 'new spirit of globalized capitalism' (Boltanski and Chiapello 2005) has affected public spaces as well. Besides the increased control over public space, its restricted access and different forms of privatization (Németh 2012), there are reverse processes at the same time: the occupation movement is only one of them, the most radical and the most publicly discussed. Simultaneously, there are many other social movements of global range with sharp social and economic messages (a good account is provided by Böhm et al. 2010), but also movements with ecological, environmentalist and sustainability concerns, such as slow cities, voluntary simplicity movements, eco cities, and many others challenging the kernel of the consummative culture: they do not want simply to restore the authentic life (to consume 'authenticity', Zukin 2008) but generally to live, to produce and consume – 'prosume' – sustainably 'according to the carrying capacity of nature' (the Aalborg Charter of European Cities and Towns towards Sustainability 1997). Thus the globalized financial capitalism, by producing global problems, also began to produce global actors to cope with them. By the mid-1980s this was explicitly stated in two different discourses: the scientific analysis of *risikogesellschaft* by Ulrich Beck (1986), and the UN report *Our Common Future*, prepared by the Brundtland Commission. What Beck described as increasing hazards in our life, permeating through borders and thus universalizing conditions, was recognized in the Brundtland report as a new global paradigm of interconnectedness between economic development, environmental degradation, and social inequality, which 'are becoming ever more interwoven locally, regionally, nationally, and globally into a seamless net of causes and effects' (World Commission on Environment and Development 1987, 13).

In a nutshell: after the alarming death of public space during the 1990s (Davis 1992, Sorkin 1992, Mitchel 1995), we now observe its revival, even radicalization. How to interpret this process? Is it a temporary release of mass discontent or a sign of deeper systemic changes in society, and what can we expect in future? To respond to these questions and to consolidate an overall view about the status of the public realm today, I will analyze the changes in the main elements preconditioning the formation of the bourgeois public sphere from early modernity, and then I will focus on key cases of contemporary public spaces serving as symbolic global signposts of social protest, trying to outline the new trends in public-realm development.

The point of departure is the presumption that the public sphere[2] today is undergoing major restructuring, similar to its transformation in France and England in the seventeenth and eighteenth centuries, revealed in the seminal work *The Structural Transformation of the Public Sphere* by Jürgen Habermas (1989). According to this understanding, public space can fulfil its functions as a node of critical discourse and societal integration only when a balance of mutual control between state, private sphere and society is established without the superdomination of any of these three. Theoretically, while the anonymous power of a bureaucratic state machine guarantees equal access to rules, the independent private sphere, consolidated by the market as a field of free entrepreneurship and competition, enables the actions of autonomous citizens, not mere subjects to the state. This also guarantees the balance between 'enlightened self-interest and orientation to common good, between the roles of clients and citizens' (Habermas 1992, 449).

These elements have been dramatically transformed today, and the balance between them disrupted – in the age of globalization, the nation-state, although still a robust structure, is losing its determining power of organizing economic and cultural principle in the face of supranational associations and alliances; the long arm of the market has become digitized and the whole economy more dependent on financial transactions than on real production relations, thus diluting the idea of 'common good'; and finally, the private sphere of autonomous citizens, capable of critical reflection, has been shattered and alienated. All this has unlocked irreversible changes in urban public spaces of contemporary cities as the last resource of the powerless populace.

Relativized space in a global world: the crawling financialization

Zygmunt Bauman's *Globalization: The Human Consequences* gives a good account of the development of our world *au fin de siècle* characterized by space unboundness, uprootedness or otherwise, deterritorialization (Appadurai 1996). Bauman describes in detail the unevenness and new injustices based on space movement: the new postmodern nomads composed of two polarized groups – the fluid global political and economic elite, and the new poors, rooted into their localities, deprived of their meaning of life. But globalization augmented the inequalities between states as well. Symptomatically, it was the mid-1990s, as documented in Bauman's book, when Bill Clinton, the spokesman for one of the most powerful elites in the world, declared publicly that for the first time the difference between internal affairs and foreign policy had ceased to exist (Bauman 1998, 13). While the political space is extended beyond national borders, human space is diminished in the cities to the secure seclusion of home: public spaces are drained of public meaning and increasingly reduced to locales of consumption and entertainment. As noted by Ileana Apostol in a study of the production of public spaces, which revealed simultaneous privatization of the public realm and alienation of the private one, 'Most of the newly developed physical public spaces are privately owned and designed to deliver pre-packaged spatial experiences that provide entertainment through surprises, distractions and sequences of events for pedestrians' (Apostol 2007, 214).

One of the most destructive aspects of globalization is the ubiquitous penetration of financial logic in all spheres of social life. As Paul Dembinski outlined in his research, finance, like all free-market economic activities, initially emerged out of the interplay between the supply of services and demand for them, but it was a long progression from 'microeconomic decisions to macroeconomic realities' (2009, 54), that gradually led to the absolute financial domination upon society and its values. For Luc Boltanski and Eve Chiapello, key elements of this process are the 'deregulation of financial markets, their de-compartmentalization, and the creation of "new financial products" that have multiplied the possibilities of purely speculative profits, whereby capital expands without taking the form of investment in productive activity' (2005, xxxvii). Dembinski describes this transformation as a 'dual shift – first from "finance as rationality" to "finance as pattern of behaviour" and then to "finance as organizing principle" – [that] leads to far-reaching psychological, social, economic and political changes' (2009, 6).

The pernicious effects of this process first disable a state which has remised a considerable amount of its social obligations to financial speculation. For example, huge liquid assets were concentrated in the hands of mutual investment funds, insurance companies and pension funds, empowering their capacity to influence the markets in accordance with their interests (Boltanski and Chiapello 2005).[3] This is a process of hidden re-articulation of societal interests which became overridden by those of the financial oligopoly and its clients.

Further on, with the personalization of marketing relationships 'the price ceases to be an objective, public factor and instead becomes a subjective, private one' (Dembinski 2009, 128), and the market ceases to be a universal regulator in the society. The sphere of labour and employment has also deteriorated. Options for hiring on a temporary basis, using a temporary workforce, flexible hours, and a reduction in the costs of lay-offs, have developed considerably in the Organisation for Economic Co-operation and Development countries, gradually whittling down the social security systems established during a century of social struggles (ibid.). The number of 'atypical jobs' (fixed-term contracts, apprentices, temps, paid trainees, beneficiaries of state-aided contracts and government-sponsored contracts in the civil service) doubled between 1985 and 1995 (Boltanski and Chiapello, xxxvii). In a study of the changing idea of success at the end of the twentieth century, the British sociologist Ray Pahl reveals the 'flexibilization' of employment expressed in the jargon of the late 1980s – *contracting out, downsizing, delayering* – denoting 'forced self-employment or consultancy on many who in the past would have settled into stable careers' (Pahl 1995, 2). In a more radical tone, Costas Douzinas discusses, in his analysis of protest movements in Athens, the new forms of employment and long periods of unemployment as a result of later capitalism which abolished permanent work and created a large number of superfluous people (Douzinas 2013, 135). This time bomb detonated into a series of mass protests and occupations in 2008 and afterwards.

Global mega-players, de-etatized states, privatized society

Another emerging kind of economic actor, the lords of financialization, reveals something about the essence of the process of financialization:

> Such players were not entrepreneurs, for they did not build or produce anything, nor were they consumers, for they did not consume. Instead, they managed value over time: intermediaries, specialized in various forms of advice, valuing, rating, guidance and assistance, or even in creating innovative investment and management tools such as investment funds.
>
> (Dembinski 2009, 46)

The constitution and concentration of global oligopolies had long-term structural implications. These mega-players are multinational firms, governments and globalized banks who emerged as winners from this redeployment of world capitalism:

> The slowdown in the world economy over almost thirty years has not really affected them, and their share of world GDP, itself rising, has continued to grow: from 17 per cent in the mid-1960s to more than 30 per cent in 1995.
>
> (Boltanski and Chiapello, xxxvii)

As Frederic Clairmont (1997) concludes, by the end of the 1990s people were moving towards one planetary government of about two hundred multinational companies who control the whole world. This is a continuous trend confirmed by numerous economic data and estimates provided by the Observatoire de la Finance: in 2005 the top 1,000 financial and industrial enterprises in terms of stock-market value accounted for 46 per cent of world capitalization, and the eight hundred largest non-financial enterprises' share of world product was about 11 per cent – equivalent to that of the 144 poorest countries. This contribution is made with the help of thirty million staff worldwide, whereas the active population in the 144 poorest countries is more than one billion – almost thirty times as many people (Dembinski 2009, 105–107). For Dembinski this difference reveals the 'huge abyss in labour productivity between the powerhouses of the world economy and the developing countries' (ibid.), but it can be interpreted also as a result of the huge difference in the profitability in the sector of financial speculations and all other economic sectors, especially that of industrial production and exploitation of raw materials.

The cited disproportion is also a sign that the classical relationship between nation-state and capital has been broken: capital flows beyond national boundaries which once served as strongholds of capitalism. If in the nineteenth and early twentieth centuries strong companies meant also strong nation-states, now this relationship is not at all so obvious: strong international companies can sometimes mean nominal, disabled, or even 'de-etatized' states, withdrawn from their basic obligations to their citizens. As Arturo Bris noted, 'It is paradoxical that, in some cases, banks and firms are so rich that they could buy entire countries' (2014). For the inhabitants of such countries, the place in which they live would probably not be able to generate an existential meaning and positive identity, but rather ambiguity and desire for emigration. The Latin-American anthropologist Nestor García Canclini, reflecting on the uncertain future of the weak nation-states of Latin-America in the 1980s and 1990s, wrote:

> Men and women increasingly feel that many of the questions proper to citizenship – where do I belong, what rights accrue to me, how can I get information, who represents my interests? – are being answered in the private realm of commodity consumption and the mass media more than in the abstract rules of democracy or collective participation in public spaces.
>
> (Thissen et al. 2013, 20)

Costas Douzinas defines this as 'biopolitical capitalism' that 'does not produce just commodities for subjects; it creates subjects. Material, social, affective, ethical and cognitive strategies are involved in this process' (Douzinas 2013, 136).

New common good: the concerns of the Risk Society

Common good is (presumably) the fuel in the motor of any concerted public action. It was exactly the common good that made a difference between the coffee houses of eighteenth-century England, one of the Habermas's favourite examples of public space, and coffee houses of the twentieth century as trivial places of consumption (Hristova 2010, 206–207). At least one of the meanings of 'public' pertains to a 'common good' or 'shared interest', which needs to be decided through discursive contestation (Fraser 1992, 128).

Is there any ground for nourishing an ideal of common good in times of recurrent crises, dissociating solidarity and the postmodern breaking apart of any common narratives? For Dembinski, financialization ruins any chance for crystallization of a common ideational project: 'in a lonely crowd of individuals linked only by transactions, the common good is an irrelevant and meaningless notion' (2009, 160). Boltanski nevertheless expresses hope (however restricted) that the 'third spirit' of globalized capitalism will create a joint perspective by connecting the idea of public good with the interests of currently dominant multinationals to preserve 'a peaceful zone at the centre of the world system, maintained as a breeding ground for cadres, where the latter can develop, raise children, and live in security' (Boltanski and Chiapello 2005, 19).

If seventeenth- and eighteenth-century bourgeois men found their common interests in modernization and consolidating nations, which was the essence of the debates in the coffee houses in England, saloons in France, and table societies in Germany – described by Habermas as genuine public spaces that gradually declined during the next two centuries – now we observe a rise in new global concerns: those of the risk society, and they are seeking expression in the public spaces of our cities.

The visions of Ulrich Beck involving the concept of 'risk society' mark the beginning of a social epoch in which several dramatic discontinuities occur. To start with, the risks of modernization ignore both class differences and national borders: 'with the endangering nature, health, nutrition and so on[,] social differences and limits are relativized'. Thus they 'display an equalizing effect among them and among those affected by them. It is precisely therein that their novel political power resides' (Beck 1997, 36).

These differences between social groups and between nation-states are still important today, as in the short run there are producers and consumers of these risks, gainers and losers. But risks of modernization sooner or later also strike those who produce or profit from them (23). In the long run risks are 'universal and supra-national', occurring around 'systematic causes' that coincide with the motor of progress and profit (40). Briefly, we all are 'living on the volcano of civilization'.

Another consequence of living in a risk society is the preoccupation with the future:

> The center of risk consciousness lies not in the presence, but in the future. [. . .] Bottlenecks in the labour market projected in mathematical models have direct effect on educational behavior. Anticipated, threatening unemployment is an essential determinant of the conditions of and attitude towards life today. The predicted destruction of the environment and the nuclear threat upset society and bring large portions of the younger generations into the streets.
>
> (Beck 1997, 40)

The transition from *class* to *risk* society is accompanied also by a value change. These are two totally different value systems because while class society is centred on the ideal of equality, the 'motive force' of risk society is safety. For Beck this is a move from the 'solidarity of need' to 'solidarity, motivated by anxiety'. The solidarity based on anticipated risks is '*negative* and *defensive*. Basically, one is no longer concerned with attaining something "good", but rather with *preventing* the worst; *self-limitation* is the goal which emerges. The *commonality of anxiety* takes the place of the *commonality of need*' (49). Sensu stricto, no 'common good' is possible any more, but only a 'common bad' that creates the ground for solidarity, stemming from anxiety, that arises and becomes a political force.

Finally, the developing risk consciousness is questioning reckless consumption, anticipating the new global threats. It is the end of the 'desire that desires desire' (Bauman 1998, 38). The endless temptations, consumers' readiness for new seduction – this is a blunt game in the face of the global crisis. Greece is an instructive example of the catastrophic results of such public policies promoting 'consumption and hedonism as the main way of linking private interests with the common good. People were treated as desiring and consuming machines. [. . .] debt-fuelled consumption was promoted as the criterion for individual happiness and social mobility' (Douzinas 2013, 136). The relentless consumerism of the 1990s, getting no satisfaction, is not possible any more. The new universal 'public bad' incites a new culture of self-limitation in personal plans, and requires new transnational forms of cooperation between public actors. *Au fin de siècle*, Habermas also joins the academic 'transnationalists', concluding that globalization of 'commerce and communication, of economic production and finance. . .and above all of ecological and military risks, poses problems that can no longer be solved within the framework of nation-states' (1998, 106).

Public space and its transfiguration in the electronic maidan

During the seventeenth and eighteenth centuries, when the early bourgeois public sphere was formed, it was a constructive element of the national project, connected with the development of printing media and channelled dissemination of information. That was the era of mercantilism, when town economies extended into national territories, a process preconditioned by the 'reading revolution' praising the printed word as a source of significant 'public' information.

During the last two decades we have witnessed simultaneously a continuation and negation of this development that took place beyond and below the increasingly porous national borders, permeating the flows of people, goods, finances and information enabled by the means of the new global connector: the internet. Society has become 'networked' (Castells 1997), 'liquid' (Bauman 2000) and 'deterritorialized' in the ethnic and media spheres, among others (Appadurai 1996), all addressing the general ontological question of the systematic consequences of the process of globalization (Thissen et al. 2013, 14), and, more particularly, the role of the world wide web in this process. When analyzing the effects of digitalization on the urban public sphere, Judith Thissen, Robert Zwijnenberg and Kitty Zijlmans draw attention to the interplay between real and digital space in terms of the experience of 'embodiment' and 'localization' (ibid., 26). The 'real' urban public sphere, although always supported by the media (newspapers, magazines, billboards, radio or television), is in terms of experience based on 'embodiment'. According to the authors, the public sphere is not so much a specific place; it consists of practices and media which to a certain extent are related to specific places.

> Their quintessence is *not* their being *localized*, but their being *embodied:* as practices of freedom, confrontation or identity construction [. . .] Sometimes the new media come rather close to *forms of embodied experience,* but, in fact, they are *essentially disembodied, virtual.*
>
> (ibid., 33)

In the next pages I will trace the process of transnationalization of public spaces through both their spatial occupation and encamping and their world-wide-web regeneration. In doing so, special focus will be given to the interplay of discursive practices of emplacement-embodiment in the city and their digital doubling; the shuttle process of 'spatial anchoring' of protests (Ramadan 2013, 146) and their return and fortification in the realm of social media. A case described by Wael Salah Fahmi provides good evidence of such shuttling when, in 1996, the Critical Art Ensemble, a group of tactical media practitioners who 'exhibit and perform at diverse venues internationally, ranging from the street, to the museum, to the internet' (www.critical-art.net/home.html) called for a strategic move away from the streets, declaring that the new geography is a virtual geography, and the core of political and cultural resistance must assert itself in electronic space, thus bringing a new model of resistant practice into action (Fahmi 2009, 23). But the reverse motion from the internet back to the street is equally indispensable: the augmented virtual sphere of a constantly growing electronic populace (Gabi and Caren 2012) will inevitably seek release in the open spaces of a city. In his analysis of the role of Twitter during the earliest days of #OccupyWallStreet discourse, Mark Tremayne says: 'Successful movements, "almost by definition", involve scale shift [. . .] This is the process by which a small or local action becomes a major social movement' (Tremayne 2014, 112).

These various practices gradually grow into a strategy involving a new kind of public spaces, transfigured in the interface between physical and digital and resulting in their doubling, extension and mutual reinforcement. The selected exemplary cases are all marked by their spatial centrality in cities, transcended into political arenas of their nation-states and beyond. They became elements of the symbolic geography of global protest which gained transnational currency through their second digitized life.

Tahrir as a proto-model of a transnational public space

There are abundant writings about the Arab Spring, Cairo protests and their generic locus, Tahrir Square, 'imagined as an auratic and magical experience' (cf. Kerton 2012; Ramadan 2013; Trombetta 2013; Fahmi 2009). The genealogy of the protest, its conception and nurturing in social media and its seemingly spontaneous organization in a process of experimental learning, are all elements of a public sphere reinvigorated 'through a political act that stages collectively the presumption of equality and affirms the ability of "the people" to self-manage and organize its affairs' (Swyngedouw, 2011a, 2011b). Tahrir Square became a central node of political struggles, an

> enclave of freedom . . . the symbol of all the social and political protests erupting in 2011 across the Mediterranean region and beyond: from Benghazi to Paternoster Square in London, from Lu Lu Square in Manama to Puerta del Sol in Madrid, and from the streets of central Athens to Rothschild Boulevard in Tel Aviv.
> (Trombetta 2013, 139)

Unlike some naïve visions about the spontaneity of the Arab Spring, a description by Lorenzo Trombetta, a Middle East correspondent for the Italian News Agency, reveals the careful preparation by activists of the January 2011 demonstrations in Cairo, and the novelty of these forms of resistance in the national framework. The young protesters used 'specific urban techniques new to the local context, by which they succeeded in taking by surprise the prevailing system of government control'. Here are some of the most typical elements of this 'new activist strategy' (Trombetta 2013, 139–141), revealed in different texts analyzing Tahrir Square, and backed up by narratives about some referential social movements which ultimately contributed to the process of the transnationalization of public space: social protests in Athens in 2008, with special focus on Syntagma Square; the Gezi movement in Istanbul since May 2013, and its evolution in Taksim Square; Occupy Wall Street; and *indignados* in Puerta del Sol in Madrid. The quotations are selected for their factual representativity and as evidence of the emotive atmosphere and revolutionary spirit in these places.

Elaborated protest strategy

- Deliberate selection of location: central places, close to the institutions of state power

'Tahrir Square was not an accidental focus for the protests; nor were the Occupy camps in Europe and North America. [. . .] the protest camp occupied the heart of Cairo's machine of power' (Ramadan 2013, 147).

- Instrumentalization of the encampment

'The act of *encampment is a vehicle* for the people to make real a new political order' (Ramadan 2013, 148).

- Mobilizing tactics

> Maximum secrecy, *new tactics for mobilizing common people* and adoption of *special rhetoric* [. . .] From a handful of young people they developed in a *seemingly spontaneous way* a few mini processions which had the objective of collecting as many people as possible and bringing them to the location of the sit-in. "I remember once, in mid-January, we *experimented with this system*: about 30 of us went in to an area, and two hours later we came out with about 1,000 people behind us".
>
> (Trombetta 2013, 140, italics added)

- Similar protest repertoire, experimenting and learning in doing

There is almost a similar repertoire of predominantly peaceful forms of protest activism across different countries and regions. The detailed descriptions by Lorenzo Trombetta (2013) contain: sit-ins; human chains of bare hands; hails of stones; protesters and policemen changing roles and adopting each other's methods; use of mobile phones as 'protest tools' for connection, coordination and the taking of photos as visual documents;[4] flexible organization and re-organization of the protesters into teams and flows; and coordination by the internet and social media. Costas Douzinas reveals similar methods used by the protesters in Athens in 2008: sit-ins, street happenings, the interruption of theatre performances and discussions with the audience, the raising of banners, occupation of a state TV studio during a broadcast and an iconic burning of a Christmas tree in Syntagma Square, plus attacks on banks and luxury shops (Douzinas 2013, 134). Similarly, the initiatives of the Gezi movement primarily revolve around public manifestations, networks of online and direct relationships, camps, discussion forums, and other meetings for debates (Farro and Demirhisar 2014, 184).

- Protest rules based on procedural equality

Special democratic rules of discussion and protests are present in the Greek and the Turkish cases. In Athens the rules stem from the traditions of the classical agora:

> In daily assemblies . . . speakers were given a number and called to the platform when their number is drawn, a reminder that most office-holders in classical Athens were selected by lot. The speakers stuck to strict 2-minute slots to allow as many as possible to contribute.
>
> (Douzinas 2013, 135)

Besides that, in weekly debates, economists, lawyers and political philosophers are invited to present

> alternatives for tackling the crisis. [. . .] These relatively new forms of resistance appear regularly now. [. . .] The occupations organize under a strict axiom of equality. Whoever is in the square, everyone and anyone, is entitled to an equal share of time to put across his or her views. The views of the unemployed and the university professor are given equal time, discussed with equal vigour and put to the vote for adoption.
>
> (ibid., 135, 138)

Similarly sensitive to procedural justice are the arrangements of the Gezi movement, prioritizing an

> organizational method that does not allow anyone to impose on others during demonstrations taking place in the internet era, that is at a time where no one can long hold positions of power in the communicative circuits (Castells 2012), which is the case of the Gezi mobilizations.
>
> (Farro and Demirhisar 2014, 184)

These procedural rules reflect deeper yearnings of protesters for social justice and equality, expressed implicitly by Douzinas: 'The right to resistance joins equality [. . .] and changes it from a conditioned norm into an unconditional axiom: people are free and equal; each counts as one in all relevant groups' (ibid.). They represent a genuine attempt to create authentic conditions for critical public debate corresponding to Habermas's ideal in which the power of arguments, and not the arguments of power (status or traditions), are involved in the decision-making.

The protesters and their social demands: regaining symbolic power

Tahrir Square attracted different protesting groups with different messages: the core group of young bloggers, educated and trained activists, and leaders of the protest had clear political demands for freedom of speech and free elections, while the poor majority were struggling for some 'basics' of life. As presented by the architect and university professor Wael Salah Fahmi, who interviewed some of the participants, they did not want any radical changes but just normal life.[5] Lorenzo Trombetta, speaking on the part of the activists, reveals a gap between them and the poor majority and their way to develop a common cause even without initially shared interests between the protesters:

> For a long time we didn't understand that people who were living on half a sandwich in the morning and the other half in the evening, were not going to be brought along to a demonstration by talking to them about the lack of freedom of speech and free elections. For this reason, instead of making use of political slogans which were incomprehensible to ordinary or poorer people, the activists managed to convince the sceptics in the poorest areas of the city with statements against the high cost of living and lack of work with phrases such as "A miserable cucumber for us, while Mubarak feasts on a whole chicken!"
>
> (Trombetta 2013, 140–141)

Often, the reports of different protest movements contain detailed analysis of participants' social background as an explanation and justification of the uprisings. In Athens these are the precariats, consisting of unemployed and unemployable young and old people, immigrants and refugees, the moving 'one-use humans' who drown in their thousands in the Mediterranean, the 'floating graveyard' of Europe. 'Getting a wage, any wage, has become the hardest quest. People beg to be exploited' (Douzinas 2013, 135). Importantly, the insurgents are conscious of their advantages in terms of their cultural and social capital: 'education and skills of networking and collaboration learnt for work' are now 'put into political practice' in Syntagma Square.

Over 60% of European youth have post-secondary education and exactly the same skills as their rulers. [. . .] One thousand unemployed lawyers, engineers and doctors are more revolutionary than 1,000 unemployed workers. The two together can change the definition of politics.

(ibid., 135–136)

In Gezi Park, unlike Syntagma Square, a small number of protesters are unemployed. The movement is constituted mainly of students and active people with liberal professions, civil servants, and other employees in the tertiary sector, with the proportion of women slightly higher than that of men. The average age is 28. 'For the participants in the Gezi movement, the economic, cultural, social, and political dimensions of individual life are closely related' (Farro and Demirhisar 2014, 178). 'Most of the Gezi movement protesters are not members of political forces or NGOs. Indeed, the majority of protesters did not relate to the "political game"' (ibid., 180).

In a similar vein, Carlos Taibo depicts the social background of Spanish *indignados*, the majority of whom, as well as being members of alternative social movements, belong to a middle class which through joblessness and insecurity has experienced an incipient process of de-classing, not so much of proletarianization (Taibo 2013, 156). Their messages drift from *ciudadanista* ('citizenism') discourse towards 'anticapitalist positions that are common in the alternative social movements' (ibid., 157). Finally, initiatives such as 15-M are, culturally and socially,

> *urban movements* [. . .] concentrated in cities with two main objectives: to prompt active decentralisation, bringing decision-making closer to the ordinary citizen; and to signal the importance of restraining the indiscriminate growth of cities and, recovering many of the already lost elements of rural life.
>
> (ibid., 158)

Identically, the social movement that became world famous under the name 'Occupy Wall Street' put a spotlight on income inequality in the US and the world. As presented by Mark Tremayne, its focus has been on '. . .skewed wealth distribution, with the top 1% owning 35% of the wealth in the USA and top 20% controlling 85%' (2014, 122).

Going back to Tahrir, the new political vigour and regained status of a central public stage transformed the square into the supreme power institution of the day, *the* 'political public space' (Lee, 2009). The year 2011 saw 'the "return of the camp" to the centre of geopolitical orderings' (Ramadan 2013, 148). It was a meaningful fact that Egypt's first democratically elected president, Mohamed Mursi, made his oath of office

> before crowds of activists and supporters packed into Tahrir Square. He declared: "*You* are the source of legitimacy and power, above all." [. . .] To swear before the camp rather than the court, before the activists rather than the judges and generals, tells us something of the central role of the protest camp in the current political moment.
>
> (ibid., 145)

The seemingly wide variety of protest subjects, ranging from urban and ecological issues to radical change of the system, are all related to two main domains: the social domain, connected to issues of employment, living conditions, security, social justice and environmental sustainability; and the political domain, pertaining to liberties and citizens' rights. Ultimately, these two domains are not independent of each other, and as the preceding analysis revealed, the problems behind the messages are structural effects of financial globalism. Therefore, the possible resolution should be of a scale and range similar to those of the causes.

Transnationalization of protests and of their public spaces

The transnational geography of Tahrir Square has unfolded through various aspects of the organization and transborder dissemination of protest 'know-how', but above all with the big media penumbra of successful revolt fascinating the global public's imagination. Notably, the name of Tahrir and its protest tactics were spread across the Middle East, through Europe and North America. The *indignados* protesters who occupied iconic public squares in Madrid and Barcelona unofficially renamed those spaces 'Tahrir Square'; during the London 'Occupy' protest a spoof street sign, 'Tahrir Square EC4M', was stuck at Paternoster Square, making the same connection (Ramadan 2013, 148; see also www.bbc.com/news/business-15326636). As clearly pointed out by Sara Fregonese (2013, 110) 'the transnationality of the uprising manifested itself in resistance practices' but also in the tools of repression used by the authorities, in both the Arab and the Western public spaces that were transformed into political arenas (Trombetta 2013, 143–144).

The recurrent occupations in Spain, Italy and Greece were inspired by Tahrir Square and in turn inspired the worldwide Occupy movement (Douzinas 2013, 135). People taking to the streets and squares across Europe, Turkey, the Middle East, Asia and the Americas (Pickerill and Krinsky 2012; Uitermark and Nicholls 2012; Teti and Mura 2013; Taibo 2013; Wallach 2013; Farro and Demirhisar 2014; Tremayne 2014) were

> against austerity, failing government policies, neoliberalism and capitalism more generally. Again, the act of encampment was a crucial tactic of protest and resistance, from Puerta del Sol in Madrid to Paternoster Square in London and Zuccotti Park in New York.
>
> (Ramadan 2013, 146)

Referring to Castells (2012), Antimo Farro and Deniz Demirhisar conclude that the practice of direct democracy adopted by different movements, such as those of the early Arab Spring (Khosrokhavar 2012), 15-M in the EU and Occupy Wall Street in the US, also represent a new challenge to institutional systems that aspire to be democratic or envision being democratized (Farro and Demirhisar, 2014, 184–185).

Sidney Tarrow defines transnational social movement as 'sustained contentious interactions with opponents – national or non-national – by connected networks of challengers organized across national boundaries' (2011, 241). The transnationalization of Tahrir Square is of a more complex character. It can be traced in all its constitutive elements: strategies and tactics of encamping, messages, operating agents, social media support and doubling in virtual reality, and finally the circulation of its image in the global symbolic iconosphere. There is also a hint of possible international support for the pre-organization of the event. Neil Staeheli and Carolyn

R. Nagel, for example, discuss the role of international civic engagement programs for the Arab Awakening, raising the question of whether the leaders of Tahrir Square who were trained by programs funded in part by the US government were promoting a Western or transnational or an 'authentic' Arab approach to citizenship and democracy (Staeheli and Nagel 2013, 116). In a way, Tahrir embodies the fragile interrelation between local and global, entangled in the mystery of whether the space occupation was an impulsive drive of the urban precariats or a result of good preliminary planning of the encamping and sit-ins, learned by the young bloggers as 'specific urban techniques', to use Trombetta's term. More importantly, the movement received global currency as its pleas became globally framed; its protest strategies recognized worldwide; and its *hacktivists*, i.e. hackers and activists (Fahmi 2009, 89), involved in transnational protest networks, have turned into 'rooted cosmopolitans' (Tarrow 2011, 238).

Certainly, social media and global communication technologies contributed to the success of the movement and its transnationalization. As reported by Ramadan, the camp was embedded in global networks of communication, with satellite news channels broadcasting 24-hour rolling coverage of the protests, and young activists tweeting and blogging from the heart of the site (2013, 147). Nevertheless, some authors, both Western and non-Western, are cautious when defining these events merely as 'Facebook and Twitter revolutions', either because this could lead to the simplistic idea that Western technologies are the true agents of liberation (ibid.), or because revolutions of the type seen in North Africa, for example, require a committed citizenry, strong bonds and effective hierarchical organizations to succeed, while Twitter and other online social networking platforms foster weak ties and horizontal decentralized organization (Tremayne 2014, 112).

However, there is no doubt that such websites have provided new ways for primarily young people to communicate, to express opinions and to receive information outside the controlled and censored spheres of national media. Besides that, this horizontal and highly accessible way of communication facilitated the logistics of the protest: the flexible organization reconvening the meeting and demonstration places; something that was necessary 'because the locations specified the day before had been heavily garrisoned by the police' (Trombetta 2013, 141). This proves the mobilization power of communication based on a new principle of organization, 'networks without leaders' (Farro and Demirhisar 2014, 181). Mark Tremayne shows, in a network analysis of the discourse in the earliest days of #OccupyWallStreet, the unexpected power of social media (Twitter) to unite and mobilize (sometimes entirely by chance) different users, groups and networks, with more or less similar ideas but with different narratives and levels of speech aggression, something that would hardly happen in 'reality' (2014, 121–122).

Besides that, there is always potential for manipulation, control, intimidation or even silencing citizens. An instructive example comes from Iran, where Facebook and Twitter as important tools for information and initial planning of protests, as in Egypt (Ramadan 2013, 147), have been suppressed by Iranian authorities, and cell phone access and text messaging were blocked during the demonstrations in 2009 (Tremayne 2014, 112).

In this process of transnationalization of public spaces of resistance, virtual and physical realities merge into 'networks of alternative communication', and a new space emerges, *Notopia*:[6] a place without location, a moving place inhabited by antiglobal hacktivists, constantly travelling and reassembling for street demonstrations in various cities worldwide. These transnational networks are linked to 'social centres' that offer their members shelter in abandoned urban spaces. This marks a profound shift from the immobility of the old protest culture of the working class to highly mobile cosmopolitan activism (Fahmi 2009, 96–97).

After the spectacle

The political radicalization brought about numerous consequences, the first and the most important of which is the transnationalization of public space. That laid the foundation of a new protest geography beyond national borders – fusing into one symbolic sequence Syntagma, Tahrir, Puerta del Sol, Paternoster Square and Zuccotti Park; uniting into one the contesting nomads: French and Belgian *Sans-Papiers*, the Spanish *indignados*, the Greek *aganaktismenoi*, the Wall Street occupiers, all these anonymous people throughout the globe.

The process of transnationalization of public space has been preconditioned by the globally universalized poverty and the new growing class of educated precariats, and technically facilitated by the internet and social media allowing coordination and synchronization of their actions. Marion Hamm, an 'ar/ctivist' as she defines herself, provides us with fresh insights into the changed spatial sensitivity of young activists of the new transnational public sphere whose practice is not simply

> a virtual reality as it was imagined in the eighties as a graphical simulation of reality. It takes place at the keyboard just as much as in the technicians' workshops, on the streets and in the temporary media centers, in tents, in socio-cultural centers and squatted houses.
>
> (Hamm 2003, 6)

Her interest in the production of emancipatory public sphere through reclaiming the streets – 'how does that work in a society [. . .] in which urban space is progressively trimmed to neoliberal/economic imperatives' (Hamm 2003, 1) – is accompanied by inquiry into the conversion of relationships from virtual to material space, and discovery of the growing relativity produced in their interaction.

Consequently, there is a new sense of place and, more generally, 'repositioning' of space in contemporary de/reterritorialized culture: it involves recognition of the importance of space by the post-consumerist homo-politicus, expressed in the spatialization of protests. In other words, the street and the square have now returned to politics (Douzinas 2013, 134). Although it is not a novelty that every revolution has its square, now 'it is an active process of intervention through which (public) space is reconfigured and through which – if successful – a new socio-spatial order is inaugurated' (Swyngedouw 2011b). This means also that public space is turned into a stake of the 'political game' in which the opponents try to outwit each other to grasp strategic hold on it.[7] In this struggle public space regains importance as a source of ultimate political legitimacy: it is well exemplified in the symbolic act of Mohamed Mursi, who, on June 29, 2012, was sworn in as President at Tahrir Square before a huge crowd of activists and supporters, where he declared: 'There is no power above people power' (http://english.alarabiya.net/articles/2012/06/29/223449.html). Soon the same power called for his resignation, and one year later he was unseated by a military council.

The name of Tahrir Square, however, remained in the mediascape as a landmark in the new transnational geography of struggles over space.

A recent example of the heightened public awareness about the importance of public space is the occupation of the Amsterdam University, described by one of the protesters, Donia Ahmadi, as 'embodied resistance to the conventional tendency to maintain dominator values to higher education'. Apart from the occupiers' topic (their rights as 'primary stakeholders in education'), the case is interesting to us because of the students' explicitly spatial strategy. Their electronic public appeal calls for 'access to university resources of which space is a basic one', and includes detailed argument referring to leftist scientists like David Harvey and bell hooks and pertaining to the right to the city beyond 'the individual liberty to access resources' and transcending to 'a common right to change ourselves by changing our city' (Ahmadi 2015).

The dialectics of place involves equally construction and deconstruction, mobilities and moorings, territorialization and deterritorialization:

> Mobilities are centrally involved in reorganizing institutions, generating climate change, moving risks and illnesses across the globe, altering travel, tourism and migration patterns. [. . .] Changes also transform the nature, scale and temporalities of families, "local" communities, public and private spaces, and the commitments people may feel to the "nation".
>
> (Hannam et al. 2006, 2)

The highly politicized public space today is an amalgam of tangible and intangible: people with their bodies and minds, webs and networks, physical and virtual places, visions and emotions, all fused in an easily explosive alloy. Among all these elements only the physical places within cities are immovable; all the other elements are flexible and volatile, moving easily from place to place, thus creating the global socio-spatial continuum of resistance. Some authors extend the argument of dynamism of places to its extreme. Mimi Sheller says:

> Places are like ships, moving around and not necessarily staying in one location. In the emerging mobilities paradigm places themselves can be seen as becoming or traveling, slowly or quickly, through greater or shorter distances and within networks of both human and non-human agents. Places are about relationships, about the placing of peoples, materials, images and the systems of difference that they perform.
>
> (2001, 13)

Of course, places literally cannot be moved but they can be constructed, reconstructed and reappropriated, with 'events, situations, moments, that temporarily disrupt the mechanisms of control and spectacle' (www.republicart.net/disc/realpublicspaces/real_editorial.htm#top).

The spectacle defined by Debord as 'the perfect image of the ruling economic order, the economic realm developing for itself' (1995, 15–16), 'turning the whole planet into a single world market' (27), has completed the 'commodity colonization of social life' (29). The possible exodus of this development is the final dehumanization of society and even its extinction. The alternative could be the restoration of the public contract and reclaiming of the public good in the framework of the 'postnational constellation'. As noted by Habermas, 'If the democratic process is to secure a basis for legitimacy beyond the nation-state, then neither state structures nor market mechanisms, but popular processes of collective will-formation alone will have to provide it' (2001, XIV). In a similar vein, Ulrich Beck proposes the idea of a new public contract based on the civic capacity for self-organization. The contract should economize the voluntary civic labour of all those invisible precariats who can make cities liveable, their energy efficient and democracy reinvigorated. Public spaces today are the legitimate place for this collective will-formation; they are the open stage for dialogue as opposed to the spectacle (Debord 1995, 17). Their changing status as transnational arenas is accompanied by shifting boundaries of our own understanding of what is 'public' in the common space and how far the right to the city extends.

Notes

1 Dembinski defines *financialization* as 'this process, which is profoundly altering the way in which the two primary elements in all societies – relationships and transactions – relate to one another. *Financialization* has a very different meaning in English-language and Marxist literature, namely the emergence of "financial capitalism" as opposed to "industrial capitalism". Some authors, such as Gerald Epstein, have used the term to mean "the increasing role of financial motives, financial markets, financial players and financial institutions in the operation of the domestic and international economies"' (2009, 9).

2 For the sake of clarity, we accept that public space is the physical representation of the public sphere, using the distinction between these terms introduced by Ali Madanipour: 'I have used the term public space (and public place) to refer to that part of the physical environment which is associated with public meanings and functions. The term public sphere (and public realm), however, has been used to refer to a much broader concept: the entire range of places, people and activities that constitute the public dimension of human social life. Therefore, public space and public sphere are not co-extensive; public space is a component part of the public sphere' (2003, Introduction, 3).

3 In a similar vein, Dembinski points out that 'over the past ten years, resources entrusted to pension funds have been growing at an average rate of almost 20 per cent a year, which is four or five times the rate of economic growth. The resources accumulated in pension funds and other social welfare institutions are estimated at fifteen trillion dollars and account for some 30–60 per cent of households' gross savings' (2009, 43).

4 '"They were not just curious passers-by. They were protestors waiting to join the demonstration," recalled Awwad. [. . .] "Instructions had gone out on the Internet not to arrive at the rallies in groups," Maher said. "In order not to attract attention, people should arrive alone and stop a short distance away from the planned location, with a newspaper to read, a book, as if they were there by chance or running an errand." [. . .] Well-prepared: It seemed like chaos but it was not. In an alley the crowd reorganized itself, handing out lemonade and Coca-Cola to wash their faces after the tear gas' (Trombetta 2013, 141–142).

5 In the best case, freedom is the last among their messages: 'We're a group of youth protesting the current conditions of no health, no education, no work, no housing, no freedom. That's why we decided that it's necessary to change all that' (Fahmi 2009, 96–97).

6 As referenced by Fahmi, *Notopia* is a term coined by Papadimitriou, and is derived from the Greek term *no-topos*, meaning 'no place, no location, no map'. See Papadimitriou (2006). In Fahmi (2009, 96–97).

7 As reported by Trombetta, '. . .both sides studied the methods of the other, readjusting their approach accordingly in an attempt to outwit each other, and using the features of the existing urban fabric of Cairo to implement their goals' (2013, 140). The stake of this outwitting is preservation of the political order or its subversion: at Tahrir Square the space has two contrary uses: as 'an instrument of sovereign power and intensified biopolitical control' and as 'a space of freedom, resistance and liberation, a space *beyond* the control of the state and outside the normal political order, in which a more progressive politics was forged and made real' (Ramadan 2013, 146).

References

Aalborg Charter (1994). www.sustainablecities.eu/fileadmin/repository/Aalborg_Charter/Aalborg_Charter_English.pdf [accessed May 11, 2017].

Ahmadi, D. (2015). The Case of the Student Occupations in Amsterdam. http://theprotocity.com/students-right-to-the-neo-liberal-university/ [accessed March 12, 2015].

Appadurai, A. (1996). *Modernity at Large: Cultural Dimensions of Globalization.* Minneapolis, MN: University of Minnesota Press.

Apostol, I. (2007). *The Production of Public Spaces: Design Dialectics and Pedagogy.* Dissertation presented to the Faculty of Graduate School, University of Southern California: UMI.

Aristotle (1962). *La politiques.* Trans by J. Tricot. Paris: Vrin.

Bauman, Z. (1998). *Globalization: The Human Consequences.* Cambridge: Polity Press.

Bauman, Z. (2000). *Liquid Modernity.* Cambridge: Polity Press.

Beck, U. (1992) [1986]. *Risk Society: Towards a New Modernity.* New Delhi: Sage. (Translated from the German *Risikogesellschaft: Auf dem Weg in eine andere Moderne.*)

Beck, U. (1999) [1997]. *What Is Globalization?* Cambridge: Polity Press.

Boltanski, L. and Chiapello, E. (2005). *The New Spirit of Capitalism.* Trans. by G. Elliott. London and New York, NY: Verso.

Böhm, S., Dinerstein, A. C. and Spicer, A. (2010). (Im)possibilities of Autonomy: Social Movements in and beyond Capital, the State and Development. *Social Movement Studies: Journal of Social, Cultural and Political Protest,* (9)1: 17–32.

Brenner, N., Marcuse, P. and Mayer, M. (2009). Cities for People, Not for Profit. *City: Analysis of Urban Trends, Culture, Theory, Policy, Action,* 13(2–3): 176–184.

Bris, A. (2014). Eight Reasons Why a New Global Financial Crisis Could Be on the Way. www.imd.org/research/challenges/TC060-14-meltdown-2015-arturo-bris.cfm [accessed February 22, 2015].

Calhoun, C. (ed) (1992). *Habermas and the Public Sphere.* Cambridge, MA: MIT Press.

Castells, M. (1997). *The Power of Identity.* Oxford and New York, NY: Blackwell.

Castells, M. (2012). *Networks of Outrage and Hope: Social Movements in the Internet Age.* Cambridge and Malden, MA: Polity Press.

Clairmont, F. (1997). Vers un gouvernement planetaire des multinationais. Ces deux cents societes qui contrôlent le monde. *Le Monde diplomatique,* April 1997. Quoted by Boltanski, L. and Chiapello, E. (2005, xlvi).

Crouch, C. (2005). *Post-Democracy.* Cambridge: Polity Press.

Davis, M. (1992). Fortress Los Angeles: The Militarization of Urban Space. In Sorkin, M. (ed), *Variations on a Theme Park: The New American City and the End of Public Space.* New York, NY: Hill and Wang.

Debord, G. (1995) [1970]. *Society of the Spectacle.* Detroit, IL: Red and Black.

Dembinski, P. H. (2009). *Finance: Servant or Deceiver? Financialization at the Crossroads.* London: Palgrave Macmillan.

Douzinas, C. (2013). Athens Rising. *European Urban and Regional Studies,* 20(1): 134–138.

Epstein, G. A. (2005). *Financialization and the World Economy.* Cheltenham: Edward Elgar Publishing.

Fahmi, W. S. (2009). Bloggers' Street Movement and the Right to the City. (Re)claiming Cairo's Real and Virtual 'Spaces of Freedom'. *Environment & Urbanization,* 21(1): 89–107.

Farro, A. and Demirhisar, D. (2014). The Gezi Park Movement: A Turkish Experience of the Twenty-First-Century Collective Movements. *International Review of Sociology: Revue Internationale de Sociologie,* 24(1): 176–189.

Fraser, N. (2014). Can Society Be Commodities All the Way Down? Post-Polanyian Reflections on Capitalist Crisis. *Economy and Society,* 43(4): 541–558.

Fraser, N. (1992). Rethinking the Public Sphere: A Contribution to the Critique of Actually Existing Democracy. In Calhoun, C. (ed), *Habermas and the Public Sphere.* Cambridge, MA: MIT Press, 109–142.

Fregonese, S. (2013). Mediterranean Geographies of Protest. *European Urban and Regional Studies*, 20(1): 109–114.

French, S., Leyshon, A. and Wainwright, T. (2011). Financializing Space, Spacing Financialization. *Progress in Human Geography*, 35(6): 798–819.

Gabi, S. and Neal, C. (2012). Occupy Online: How Cute Old Men and Malcolm X Recruited 400,000 US Users to OWS on Facebook. *Social Movement Studies: Journal of Social, Cultural and Political Protest*, 11(3–4): 367–374.

Habermas, J. (1989) [1962]. *The Structural Transformation of the Public Sphere: An Inquiry into a Category of Bourgeois Society*. Cambridge, MA: Polity Press.

Habermas, J. (1992). Further Reflections on the Public Sphere. In Calhoun, C. (ed), *Habermas and the Public Sphere*. Cambridge, MA: MIT Press.

Habermas, J. (1998). *The Inclusion of the Other: Studies in Political Theory*. Cambridge, MA: Polity Press.

Habermas, J. (2001). *The Postnational Constellation*. Cambridge, MA: Polity Press.

Hannam, K., Sheller, M. and Urry, J. (2006). Editorial: Mobilities, Immobilities and Moorings. *Mobilities*, 1(1): 1–22.

Hamm, M. (2003). *A r/c tivism in Physical and Virtual Spaces*. http://republicart.net/disc/real publicspaces/hamm02_en.pdf [accessed June 16, 2014].

Hristova, S. (2010). Imagining the City as a Space for Cultural Policy. *Sociological Problems*, special issue by the Bulgarian Sociological Association, 'Sociology on the Move', devoted to XVII ISA World Congress: 199–221.

Kerton, S. (2012). Tahrir, Here? The Influence of the Arab Uprisings on the Emergence of Occupy. *Social Movement Studies: Journal of Social, Cultural and Political Protest*, 11(3–4): 302–308.

Khosrokhavar, F. (2012). *The New Arab Revolutions that Shook the World*. Boulder, CO: Paradigm Publishers.

Lee, N. K. (2009). How is a Political Public Space Made? The Birth of Tiananmen Square and the May Fourth Movement. *Political Geography*, 28(1): 32–43. Cited after Ramadan 2013: 148.

Lefebvre, H. (2009). [1966] Theoretical Problems of Autogestion. In Brenner, N. and Elden, S. (eds), *Henri Lefebvre: State, Space, World*. Minneapolis, MN: University of Minnesota Press, 138–152.

Madanipour, A. (2003). *Public and Private Spaces of the City*. New York, NY: Routledge.

Mitchell, D. (1995). The End of Public Space? People's Park, Definitions of the Public, and Democracy. *Annals of the Association of American Geographers*, 85(1): 108–133. http://links.jstor.org/sici?sici=0004-5608%28199503%2985%3A1%3C108%3ATEOPSP%3E2.0.CO%3B2-M [accessed May 20, 2012].

Moreno, L. (2014). The Urban Process Under Financialised Capitalism. *City: Analysis of Urban Trends, Culture, Theory, Policy, Action*, 18(3): 244–268.

Németh, J. (2012). Controlling the Commons: How Public is Public Space? *Urban Affairs Review*, 48(6): 811–835.

Pahl, R. (1995). *After Success: Fin-de-Siecle Anxiety and Identity*. Cambridge, MA: Polity Press.

Papadimitriou, F. (2006). A Geography of 'Notopia': Hackers et al., Hacktivism, Urban Cybergroups/Cyber-Cultures and Digital Social Movements. *City: Analysis of Urban Trends, Culture, Theory, Policy, Action*, 10(3): 317–326.

Pickerill, J. and Krinsky, J. (2012). Why Does Occupy Matter? *Social Movement Studies: Journal of Social, Cultural and Political Protest*, 11(3–4): 279–287.

Ramadan, A. (2013). From Tahrir to the World: The Camp as a Political Public Space. *European Urban and Regional Studies*, 20(1): 145–149.

Rancière, J. (1999). *Disagreement: Politics and Philosophy*. Minneapolis, MN: University of Minnesota Press.

Sassen, S. (2012). The Global Street or the Democracy of the Powerless. Interview by Łukasz Pawłowski in *Kultura Liberalna*, *163, 8/2012*. Available at: http://kulturaliberalna.pl/2012/02/20/the-global-street-or-the-democracy-of-the-powerless/ [accessed March 19, 2015].

Sheller, M. (2001). The Mechanisms of Mobility and Liquidity: Re-thinking the Movement in Social Movements. Lancaster: Department of Sociology, Lancaster University. http://www.lancaster.ac.uk/fass/resources/sociology-online-papers/papers/sheller-mechanisms-of-mobility-and-liquidity.pdf [accessed May 11, 2017].

Sorkin, M. (ed) (1992). *Variations on a Theme Park: The New American City and the End of Public Space*. New York, NY: Hill & Wang.

Staeheli, L. and Nagel, C. R. (2013). Whose Awakening is It? Youth and the Geopolitics of Civic Engagement in the 'Arab Awakening'. *European Urban and Regional Studies*, 20(1): 115–119.

Swyngedouw, E. (2011a). Interrogating Post-Democratization: Reclaiming Egalitarian Political Spaces. *Political Geography*, 30(7): 370–380.

Swyngedouw, E. (2011b). Every Revolution has its Square. http://citiesmcr.wordpress.com/2011/03/18/every-revolution-has-its-square/ [accessed December 16, 2014].

Taibo, C. (2013). The Spanish *Indignados*: A Movement with Two Souls. *European Urban and Regional Studies*, 20(1): 155–158.

Tarrow, S. (3rd ed. 2011). *Power in Movement: Social Movements and Contentious Politics*. Cambridge, MA: Cambridge University Press.

Teti, A. and Mura, A. (2013). Convergent (Il)liberalism in the Mediterranean? Some Notes on Egyptian (Post-)authoritarianism and Italian (Post-)democracy. *European Urban and Regional Studies*, 20(1): 120–127.

Thévenot, L. (2014). Voicing Concern and Difference: From Public Spaces to Common-Places. *European Journal of Cultural and Political Sociology*, 1(1): 7–34.

Thissen, J., Zwijnenberg, R. and Zijlmans, K. (eds) (2013). *Contemporary Culture: New Directions in Arts and Humanities Research*. Amsterdam: Amsterdam University Press.

Tremayne, M. (2014). Anatomy of Protest in the Digital Era: A Network Analysis of Twitter and Occupy Wall Street. *Social Movement Studies*, 13(1): 110–126.

Trombetta, L. (2013). More than Just a Battleground: Cairo's Urban Space During the 2011 Protests. *European Urban and Regional Studies*, 20(1): 139–144.

Uitermark, J. and Nicholls, W. (2012). How Local Networks Shape a Global Movement: Comparing Occupy in Amsterdam and Los Angeles. *Social Movement Studies: Journal of Social, Cultural and Political Protest*, 11(3–4): 295–301.

Vicari, S. (2014). Networks of Contention: The Shape of Online Transnationalism in Early Twenty-First Century Social Movement Coalitions. *Social Movement Studies: Journal of Social, Cultural and Political Protest*, 13(1): 92–109.

Wallach, Y. (2013). The Politics of Non-iconic Space: Sushi, Shisha, and a Civic Promise in the 2011 Summer Protests in Israel. *European Urban and Regional Studies*, 20(1): 150–154.

World Commission on Environment and Development (1987). *Our Common Future*. Report of the Brundtland Commission. Oxford: Oxford University Press.

Zukin, S. (2008). Consuming Authenticity: From Outposts of Difference to Means of Exclusion. *Cultural Studies*, 22(5): 724–748.

3 Seeing the local in global cities

Jerome Krase

Introduction

This chapter presents a visual sociological approach that I have taken in a long series of studies of urban neighbourhoods. A brief review of some of the most important theoretical perspectives on these interrelated phenomena, such as those of Saskia Sassen, David Harvey, and Manuel Castells, isolates common expectations about the visibility of competing spatial practices in shared multi-ethnic residential and commercial environments. It is argued that the many of the contradictions created by the concentration of global capital can be seen in the streetscapes of immigrant and migrant neighbourhoods of global cities. From Georg Simmel, through Henri Lefebvre, and Lyn H. Lofland, the visible and the symbolic have been central to urban analysis. Therefore, glocalization is addressed here by focusing on ubiquitous aspects of what Jackson (1964) called 'vernacular landscapes,' such as commercial signs and graffiti that can be seen as local markers of the process of globalization. After a short treatment of visual sociology, the chapter concludes with a small selection of photographs taken by the author in Berlin, Frankfurt am Main, London, New York, and Rome. The images are discussed as they reflect the symbolic competition between more or less recent migrants to claim 'contested terrains.'

It is ironic that, for those who study cities, global forces have increased the need to understand the meanings of local streetscapes. The eyes of many researchers have refocused on the competition between migrant identities and how their spatial practices are indicators of globalized or, perhaps more accurately, 'glocalized' cities; how the everyday lives of city dwellers represented in the built environment have long been of interest to architects and planners who study vernacular landscapes. The indigenous practices of ordinary people who have been brought together by distant forces and processes are small-scale examples of that for which Roland Robertson (1997) coined the term *glocalization*. According to Juergen Osterhammel and Niels P. Petersson, 'Robertson recognized that homogenizing and universalizing forces of globalization do not obliterate the heterogeneity and particularity of local forces as much as their interaction creates varying degrees of hybrid culture' (Osterhammel and Petersson 2005, 7).

Georg Simmel early on established the central role of the visible in theorizing about the complex and constantly changing metropolis. This continues as a tradition in all the urban sciences, if only as a powerful subtext. A century ago he wrote that

Modern social life increases in ever growing degree the role of mere visual impression which always characterizes the preponderant part of all sense relationships between man and man, and must place social attitudes and feelings upon an entirely changed basis.

(Simmel [1908] 1924, 360)

My own synthesis of theories about visualizing spatial practices, applied in a number of different places, is simply that ordinary people change the meaning of spaces and places by changing their appearance (Krase 1993, 2002, 2004a). For example, as I had also written about Italian neighbourhoods in the United States,

> Beyond the great public spaces and edifices lies a vast domain of little people and little structures which in fact comprise most of our material society and where ordinary people have created distinct landscapes and places. The designs of these neighbourhoods are such in the way that space is socially constructed. Italians, like all migrants, carry designs or living from the original home environments and adapt them to the resources and opportunities in new locales.
>
> (Krase 2004b, 27)

This work is also greatly informed by that of Lyn H. Lofland (1998; 2003, 938–939), and the theoretical perspective of Symbolic Interactionism. She had noted that interactionists have contributed to urban studies by showing how people communicate through the built environment, for example, seeing settlement as symbol. Individuals and groups interact with each other through visual images by effecting what it is that people see on the streets. The meanings of what they see, however, come from a different socialization source. Lofland (1985, 22) also argued that 'the city, because of its size, is the locus of a peculiar social situation; the people found within its boundaries at any given moment know nothing personally about the vast majority of others with whom they share this space.' She adds that 'city life was made possible by an "ordering" of the urban populace in terms of appearance and spatial location such that those within the city could know a great deal about one another by simply looking.'

Global cities

Global cities are paradigmatic sites for visual and symbolic competition, and as noted by Saskia Sassen they are sites for the contradictions of the globalization of capital. The powerful as well as the disadvantaged are concentrated therein and the marginalized also find ways to claim 'contested terrain.' The global city heightens diversity by concentrating migrants and immigrants. Sassen notes that although the dominant, corporate culture, 'inscribes noncorporate cultures and identities with "otherness", thereby devaluing them, they are present everywhere.' Immigrant communities and their informal economies are common examples of this process (Sassen 2001). Tarry Hum has also written on the role of immigrant business in New York City, where she notes that while its global stature relies on nostalgia about historic immigrant enclaves, today immigrants are transforming historic landscapes not only by forming new enclaves but by the creation, and re-creation, of many multi-ethnic, multi-racial neighbourhoods (Hum 2004; Krase and Hum 2007).

Kieran Bonner closely examined Sassen's (2000) ideas about the recovery of space and place in analyses of the global economy by looking at both ends of the globalization scale, or the upper as well as the 'lower' circuits of globalization. It is at the lower circuit that one sees the multicultural world produced and one can grasp that '. . .globalization is not just about the movement of capital by global corporations, but also about the movement of people who are often in contest with such economic developments' (Sassen 2000, 277). John Brinckerhoff Jackson had also called us to look at what '. . .lies underneath below the symbols of permanent power expressed in the "Political Landscape"' (Jackson 1964, 6), because what ordinary people do in a particular physical territory and how they use objects therein are critical for understanding the space. As to why the study of vernacular as opposed to 'polite' architecture has become more valuable for insight into social history, he argued that since the nineteenth century, 'Innumerable new forms have evolved, not only in our public existence – such as the factory, the shopping centre, the gas station, and so on – but in our private lives as well' (Jackson 1964, 118–19).

As to the importance of seeing difference in the city, Richard Sennett commented on the management of difference in New York City:

> What is characteristic of our city building is to wall off the differences between people, assuming that these differences are more likely to be mutually threatening than mutually stimulating. What we make in the urban realm are therefore bland, neutralizing spaces which remove the threat of social contact: street walls faced in sheets of plate glass, highways that cut off poor neighbourhoods from the rest of the city, dormitory housing developments.
>
> (Sennett 1990, xii)

Michael Sorkin (1992, xi–xv) sees the theming of the cityscapes as an effort at social control, and laments that the consequence is the potential loss of the 'familiar spaces of traditional cities, the streets and squares, courtyards and parks', which for him 'are our great scenes of the civic, visible and accessible, our binding agents.'

The modern global city is often referred to as 'postmodern,' and despite the fact that postmodern urbanists tend to '. . .portray the contemporary city as fragmented, partitioned, and precarious, and as a result, less legible than its modernist precursor' (Beauregard and Haila 2000, 23), Robert A. Beauregard and Anne Haila believe that a distinctly postmodern city has not displaced the modern one. Rather they find there is a more complex patterning of old and new, and of continuing trends and new forces that result in different kinds of segregation and different logics of location. Especially important is the uneven spatial competition that lower-class immigrants face with more privileged members of society. In this regard, Roland van Kempen and Peter Marcuse (1997, 4) also argued that uniform patterns cannot be expected and they offer contemporary residential community forms in the 'citadels of the rich', gentrified areas, middle-class suburbs, tenement areas, ethnic enclaves, and what is to them a 'new type' of ghetto. Anthony King (1996) speaks of cities as 'text' to be read. Vernacular landscapes are crucial to that reading. Sharon Zukin also noted that the emphasis of urbanists had been on competition over access and representations of the urban centre. 'Visual artefacts of material culture and political economy thus reinforce – or comment on – social structure. By making social rules "legible" they represent the city' (Zukin 1996, 44).

Manuel Castells (1989 and 1996) looked at contested real and imagined urban spaces. For him power is information, and 'Spaces of Places' are superseded by networks of information or 'Spaces of Flows,' which leads to the tribalization of local communities. As local identities lose meaning, place-based societies and cultures, such as neighbourhoods, also lose power. To reverse the trend toward disempowerment, Castells offered reconstruction of place-based meaning via social and spatial projects at cultural, economic, and political levels. Territorially defined ethnic groups, for example, can preserve their identities and build on their historical roots by the 'symbolic marking of places,' the preservation of 'symbols of recognition,' and the 'expression of collective memory in actual practices of communication.'

Visual sociology

In discussing Henri Lefebvre's 'Spatial Practices' David Harvey noted that those who have the power to command and produce space are therefore able to reproduce and enhance their own power. It is within the parameters outlined by these practices that the local lives of ordinary urban dwellers take place. For Harvey (1989, 265),

> Different classes construct their sense of territory and community in radically different ways. This elemental fact is often overlooked by those theorists who presume a priori that there is some ideal-typical and universal tendency for all human beings to construct a human community of roughly similar sort, no matter what the political or economic circumstances.

Pierre Bourdieu (1977, 188)[1] notes that the production of such symbolic capital also serves ideological functions, because the mechanisms through which it contributes 'to the reproduction of the established order and to the perpetuation of domination remain hidden.' For visual sociology some of these 'hidden' reproductions *cum* re-presentations are in 'plain view.' People show themselves to each other in the course of their everyday lives. Bourdieu's notion of the 'habitus' or practices that produce, in this case, visible regularities is also helpful in this regard (1977, 72–95).

For Lefebvre the visual was central to the production and reproduction of social space of any scale:

> Thus space is undoubtedly produced even when the scale is not that of major highways, airports or public works. A further important aspect of spaces of this kind is their increasingly pronounced visual character. They are made with the visible in mind; the visibility of people and things, of spaces and of whatever is contained by them. The predominance of visualization (more important than "spectacularization", which is in any case subsumed by it) serves to conceal repetitiveness. People *look*, and take sight, take seeing, for life itself. We build on the basis of papers and plans. We buy on the basis of images. Sight and seeing, which in the Western tradition once epitomized intelligibility, have turned into a trap: the means whereby, in social spaces, diversity may be simulated and a travesty of enlightenment and intelligibility ensconced under the sign of transparency.
>
> (Lefebvre 1991, 75–76)

For Howard Becker (1995) as well as Carol A. B. Warren and Tracy X. Karner (2005), visual methods are important tools in the repertoire of contemporary qualitative researchers. Jon Prosser (1998, 29) argued that 'Taken cumulatively images are signifiers of a culture; taken individually they are artefacts that provide us with very particular information about our existence.' Photography is an especially valuable tool for qualitative researchers as it results in the creation of a 'different order of data, and, more importantly, an alternative to the way we have perceived data in the past' (ibid.). John Grady (1996, 14) describes it is an organized attempt to investigate 'how sight and vision helps construct social organization and meaning and how images and imagery can both inform and be used to manage social relations.' Douglas Harper, commenting on Howard S. Becker (1974), John Grady (1996) and other seminal pieces in the establishment of the field of visual sociology, extends the vision of its scope, taking into account postmodern and other critiques, but at base he argues:

> Visual sociology should, I think, begin with traditional assumptions of sociological field work and sociological analysis. The photograph can be thought of as 'data'; in fact the unique character of photographic images force us to rethink many of our assumptions about how we move from observation to analysis in all forms of sociological research. But note that I suggested that image making an analysis begins with these and other traditional assumptions and practices. It does not end there!
>
> (Harper 1998, 34–35)

Visual sociology can help to demonstrate the agency of even the least empowered through visual modification of the spaces they use and occupy.

Insights from the visual field

Now we will turn to a short series of photographs taken by the author in order to display the irony of globalization expressed in glocalized, hybridized cityscapes. They have been taken from a large collection of visual sociological research conducted on how the agency of ordinary people has given meaning to urban spaces by effecting what those spaces look like. Although there are many other 'global cities,' here we will focus on the vernacular landscapes found in the complex streetscapes of Berlin, Frankfurt am Main, London, New York, and Rome. In these five cities, more or less recent migrants, consciously and unconsciously, claim various 'contested terrains' via their spatial practices which have become common indications of globalized or, perhaps more accurately, 'glocalized' urban status. At the glocal level the most often recognized competition takes place between migrants and those already in the neighbourhood, but the rapidly changing environment creates other opportunities for competition such as that between different migrant groups (Krase 1997, 2003, 2005, 2006).

London and its greater metropolitan area are well known for neighbourhoods defined by class and ethnicity. One less known area undergoing simultaneous gentrification and ethnic influx is the central London district of Islington. In Figure 3.1 we look at a business offering *Servicio de Encomiendas*, La Tranquera restaurant and Halal chicken on a section of Holloway Road which one local resident jokingly referred to as 'Little Ecuador.' Competing for clients along the thoroughfare are many other Latino food purveyors (El Rincon de don Pepe, Quiteno, Colombia Tienda Tropicale, Café La Paz, et al.) along with eating establishments catering to Afro-Caribbean diners, South

Figure 3.1 'Latino' businesses in Islington, London, 2007

Asians, and other Londoners. Across the street from 'Little Ecuador' in this changing neighbourhood is a local chain store grocery that has responded to recent Polish migrants and workers with small handwritten signs in Polish offering Polish medicine, bread, and other products. It must be noted that unless you can read Polish, the sign would have little value except to say that this store is trying to serve people who speak a language different from English.

Many Londoners think of Hammersmith's King Street as the commercial centre for the Polish community who have been settling there at least since the end of the First World War. The Polish ethnic claim to the territory is made in various ways along the busy commercial street but especially in the impressive Polish Social and Cultural Centre. Despite the historical dominance of the area by Polish residents and businesses, it has become in recent years a highly competitive multicultural environment, especially as property values rise in London's inner suburbs. In Figure 3.2, Polish casual workers await employment opportunities on the corner near the Polish specialities food store, which is not very far away from the Thai Smile Supermarket Market and other ethnic Asian establishments that cater to Chinese, Filipino, Japanese, Malaysian, and Korean customers, among others. As noted by Ayona Datta (2009), the exposure of Polish workers to this diversity is for many a very new experience and part of their socialization into cosmopolitanism. It should also be noted that the appearance of migrant casual workers in many global cities has often been greeted with hostility, and sometimes with violence.

In New York City, as in other global cities, immigrant and migrant ethnic groups are often attracted to the same localities. The Belmont section of the New York City

Figure 3.2 Polish delicatessen and casual workers on King Street in Hammersmith, London, 2007

Figure 3.3 Mexican versus Puerto Rican graffito, Belmont, the Bronx, New York, 2006

borough of the Bronx has attracted immigrants for more than a century. Between 1920 and 1980 it had been dominated by Italians so much so that it was referred to as a 'Little Italy.' Despite the fact that those in the area who claim Italian heritage are very few today, to attract customers local businessmen claim that they are still in the 'Heart of Little Italy.' For decades Italians had uneasily co-existed with Puerto Rican residents. As shown by the graffito in Figure 3.3, for Latinos at least, the hegemonic conflict today is between Mexicans and Puerto Ricans. To make the ethnic definition of the community even more problematic, many of the 'Italian' restaurants in 'Little Italy' are owned and run by immigrants from Albania as well as Kosovo. Not far from Little Italy is a growing Albanian ethnic commercial centre which has its own distinctive visual flavour.

For both good and for bad, Kreuzberg is one of the best-known parts of Berlin, with nightlife, criminality, the drug scene, and immigrants. More recently it has attracted students, young professionals, and young couples whose visual claims on the streetscapes in the eastern part of the borough compete with its almost 'oriental' appearance. As an indicator of the competition between the alternative crowds and Turkish residents, we see in Figure 3.4 a New York subway graffito-style 'Brooklyn, N.Y.' establishment next door to a genuinely Turkish barber (Berber) shop. The well-known large open-air flea market in Kreuzberg also offers to diverse vendors an even wider variety of shoppers representing the ethnic and class diversity of increasingly cosmopolitan Kreuzberg. While Kreuzberg

Figure 3.4 Brooklyn and Berber in Kreuzberg, Berlin, Germany, 2001

Figure 3.5 Oriental *Lebensmittel*, Frankfurt am Main, Germany, 2005

thrives on its diverse cultures and is still an attractive area for the younger, alter-native type of person, the district is also characterized by high levels of structural unemployment, and income levels are among the poorest of Berlin, despite increas-ing gentrification.

In many cities in Europe one can find transient and settled immigrant areas near central train stations. Two of those which I have found to offer the greatest mix and visual competition can be found around the central train stations of the Europe Banking Center of Frankfurt am Main, and the Italian capital city of Rome. Even the briefest of strolls along the streets adjacent to these busy transportation hubs will bring pedestrians into visual contact with both non-European and European migrants who are seeking employment and other opportunities in their new home. The outdoor newspaper racks found in both of these busy multi-ethnic residential and commer-cial neighbourhoods near the central stations are also good indicators of diversity.

Figure 3.6 Chinese clothing shops, L'Esquilino, Rome, Italy, 1998

For example, between the Frankfurt am Main Hauptbahnhof and the Main River newspaper racks offer Turkish, Italian, Albanian, Polish, and other non-German language newspapers in addition to a few 'German' ones. In Figure 3.5 above we can see in the same frame some of the most well-known markers of Turkishness and Islam in Germany; *Döner*[2] and the less noticeable *Moschee*[3] sign on the busiest immigrant shopping street near the Hauptbahnhof in Frankfurt. Although, as indicated by the recent controversy over the building of a large mosque in Cologne (Landler, 2007), visible expressions of Islam are not popular sights in Germany, *Döner* has become almost a 'German' food.

Above is a photograph taken in the Esquilino neighbourhood near the Stazione Centrale in Rome which juxtaposes a more recent grocery offering Indian products and its adjacent, longer term, Chinese-owned clothing shop, one of many for which L'Esquilino has become known for shoppers. As with many other municipalities in Italy, local Roman authorities have responded aggressively and negatively to the visible claims of immigrants to venerated central city spaces. For example, in 2004 the mayor of Rome declared that 'There cannot be a Chinatown in Rome' (Williams, 2004). Despite the opposition, not only has the Chinatown not become invisible, but it has become enlivened by the addition of several other ethnic variations such as Bangladeshi and Middle Eastern jewellers. Throughout the area there are numerous, and aesthetically annoying, flyers pasted on building walls which are a common way for people in Italy to advertise, but in L'Esquilino they appear, as do commercial store signs, in several Asian languages. Therefore, they can also be taken as indicators of who is challenging whom for visual control of the 'contested' neighbourhood. (See Figure 3.7 below.)

Figure 3.7 South Asian and Italian signage, L'Esquilino, Rome, Italy, 2003

Summary

As part of an interdisciplinary dialogue, this chapter tried to demonstrate the usefulness of a visual perspective to understanding spatial forms and practices in our globalizing world. Spatialized notions of home and migration were addressed via images that reflect the symbolic competition created by more or less recent migrants as they lay claim to 'contested terrains' in five global cities. While it is true that that globalization has reduced the fixedness of 'home', it is no less important in today's cities. In the process of creating new, perhaps transnational, homes, migrants are changing the appearance and meaning of urban spaces. At the same time, their spatial practices are also challenging the dominant culture as to the definition of these contested terrains. It is important for public authorities as well as ordinary residents of global cities to recognize the 'visual contribution' to these competitions and conflicts. Finally, it is hoped that this modest exercise reveals another, perhaps 'hidden' dimension of the complex relationship between globalization and glocalization.

For support of the photographic research, acknowledgement is made here to the Rector's Committee for Scientific Research, and the Department of Sociology, University of Rome, La Sapienza; the PSC/CUNY Travel Fund; Brooklyn College Foundation; Murray Koppelman Travel Grant; and the Macaulay Honors College Travel Fund.

Notes

1 See also King (1996, 112–136).
2 *Döner* means 'revolving' in Turkish and it refers to the way the meat is grilled for the preparation of a special kind of popular snack, which has turned into a foodscape of Turkishness.
3 German for *Mosque*.

References

Beauregard, R. A. and Haila, A. (2000). The Unavoidable Continuities of the City. In Marcuse, P. and Van Kempen, R. (eds), *Globalizing Cities: A New Spatial Order?* Oxford: Blackwell, 22–36.

Becker, H. S. (1974). Photography and Sociology. *Studies in the Anthropology of Visual Communication*, 1(1): 3–26.

Becker, H. S. (1995). Visual Sociology, Documentary Photography, and Photojournalism: It's (Almost) All a Matter of Context, *Visual Sociology* 10(1–2), 5–14.

Bonner, K. (2007). Reflexive Theorizing While Travelling through Montreal and Toronto: The Global Cities Discourse, New Urbanism and the Travel Essays of Jan Morris. In Sloan, J. (ed), *Urban Enigmas: Montreal, Toronto, and the Problem of Comparing Cities*. Montreal and Kingston: McGill-Queens University Press, 274– 298.

Bourdieu, P. (1977). *Outline of a Theory of Practice*. New York, NY: Cambridge University Press.

Castells, M. (1989). *The Informational City*. Oxford: Blackwell.

Castells, M. (1996). Conclusion: The Reconstruction of Social Meaning in the Space of Flows. In Legates, R. T. and Stout, F. (eds), *The City Reader*. London: Routledge, 494–498.

Datta, A. (2009). Places of Everyday Cosmopolitanisms: East-European Construction Workers in London. *Environment and Planning A*, 41(2): 353–370.

Grady, J. (1996). The Scope of Visual Sociology. *Visual Sociology* 11(2): 10–24.

Harper, D. (1988). Visual Sociology: Expanding Sociological Vision. *American Sociologist*, 19(10): 54–70.

Harvey, D. (1989). *The Urban Experience*. Baltimore, ML: Johns Hopkins University Press.

Hum, T. (2004). Immigrant Global Neighborhoods in New York City. In Krase, J. and Hutchison, R. (eds), *Race and Ethnicity in New York City*. London: Elsevier Press, 25–55.

Jackson, J. B. (1964). *Discovering the Vernacular Landscape*. New Haven: Yale University Press.

King, A. (ed) (1996). *Re-Presenting the City: Ethnicity, Capital and Culture in the Twenty-First Century Metropolis*. London: Macmillan.

Krase, J. (1993). Traces of Home. *Places: A Quarterly Journal of Environmental Design*, 8(4): 46–55.

Krase, J. (1997). Polish and Italian Vernacular Landscapes in Brooklyn. *Polish American Studies*, 54(1): 9–31.

Krase, J. (2002). Navigating Ethnic Vernacular Landscapes Then and Now. *Journal of Architecture and Planning Research*, 19(4): 274–281.

Krase, J. (2003). Chinatown: A Visual Approach to Ethnic Spectacles. *CUNY Bulletin of Asian American/Asian Affairs*. Asian American/Asian Research Institute of the City University of New York, NY: 89–90. www.aaari.info/2003workshop4b.html [accessed February 5, 2014].

Krase, J. (2004a). Visualizing Ethnic Vernacular Landscapes. In Krase, J. and Hutchison, R. (eds), *Race and Ethnicity in New York City*. London: Elsevier Press, 1–24.

Krase, J. (2004b). Italian American Urban Landscapes: Images of Social and Cultural Capital. *Italian Americana*, 22(1): 17–44.

Krase, J. (2004c). Seeing Community in a Multicultural Society: Theory and Practice. In *Perspectives of Multiculturalism: Western and Transitional Countries*. Zagreb: Croatian Commission for UNESCO, FF Press: 151–177. http://unesdoc.unesco.org/images/0013/001375/137520e.pdf [accessed on April 19, 2014].

Krase, J. (2004d). Little Italy, identita e semiotica spaziale. In Ceramella, N. and Massara, G. (eds), *Merica: forme della cultura italoamericana*. Isernia: Cosmo Iannone, 115–142.

Krase, J. (2005). Poland and Polonia: Gentrification as Ethnic Aesthetic Practice and Migratory Process. In Atkinson, R. and Bridges, G. (eds), *Gentrification in Global Perspective*. London: Routledge, 185–208.

Krase, J. (2006). Visualizing Ethnic Vernacular Landscapes in American Cities. In Aaron, M., McCright, A. M. and Clark, T. N. (eds), *Community and Ecology: Dynamics of Place, Sustainability, and Politics*. London: Elsevier Press, 63–84.

Krase, J. (2006). Seeing Ethnic Succession in Little Italy: Change Despite Resistance. *Modern Italy*, 11(1): 79–95.

Krase, J. (2007). Seeing Ethnic Succession in Little and Big Italy. In Meir, L. and Frers, L. (eds), *Encountering Urban Places: Visual and Material Performances in the City*. Aldershot: Ashgate, 97–118.

Krase, J. and Hum, T. (2007). Ethnic Crossroads: Toward a Theory of Immigrant Global Neighborhoods. In Hutchinson, R. and Krase, J. (eds), *Ethnic Landscapes in an Urban World*. London: Elsevier Press, 97–119.

Krase, J. (2007). Visualisation du changement urbain. *Société*, 1(95): 65–87.

Landler, M. (2007). Germans Split Over a Mosque and the Role of Islam. *The New York Times*, July 5.

Lefebvre, H. (1991). *The Production of Space*. Trans. by D. Nicholson-Smith. Oxford: Blackwell.

Lofland, L. H. (1985). *A World of Strangers: Order and Action in Urban Public Spaces*. Prospect Heights, IL: Waveland Press.

Lofland, L. H. (1998). *The Public Realm: Exploring the City's Quintessential Social Territory*. New York, NY: Aldine de Gruyter.

Lofland, L. H. (2003). Community and Urban Life. In Reynolds, L. T. and Herman-Kinnery, N. J. (eds), *Handbook of Symbolic Interactionism*. Lanham, MD: AltaMira, 937–974.

Osterhammel, J. and Petersson, M. P. (2005). *Globalization: A Short History*. Trans. by D. Geyer. Princeton, NJ: Princeton University Press.

Prosser, J. (1998). The Status of Image-Based Research. In Prosser, J. (ed), *Image-Based Research: A Sourcebook for Qualitative Researchers*. London: Falmer Press, 97–112.

Robertson, R. (1997). Comments on the 'Global Triad' and 'Glocalization'. In Nobutaka, I. (ed), *Globalization and Indigenous Culture*. Institute for Japanese Culture and Classics, Kokugakuin University. www2.kokugakuin.ac.jp/ijcc/wp/global/index.html [accessed October 30, 2007].

Sassen, S. (2000). The Global City: Strategic Site/New Frontier. In Smith, M. P. (ed), *Transnational Urbanism: Locating Globalization*. Malden, MA: Blackwell: 48–61.

Sassen, S. (2001). *The Global City: Strategic Site/New Frontier*. www.india-seminar.com/2001/503/503%20saskia%20sassen.htm [accessed August 7, 2014].

Sassen, S. (2002). Globalization and Its Discontents. In Bridge, G. and Watson, S. (eds), *The Blackwell City Reader*. Oxford: Blackwell: 161–70.

Sennett, R. (1990). *The Conscience of the Eye*. New York, NY: W.W. Norton & Company.

Simmel, G. [1908] (1924). Sociology of the Senses: Visual Interaction. In Park, R. E. and Burgess, E. W. (eds), *Introduction to the Science of Sociology*. Chicago, IL: University of Chicago Press, 356–361.

Sorkin, M. (ed) (1992). *Variations on a Theme Park*. New York, NY: Hill and Wang.

Van Kempen, R. and Marcuse, P. (1997). The Changing Spatial Order in Cities. *American Behavioral Scientist*, 41(3): 285–298.

Warren, C. A. B. and Karner, T. X. (2005). *Discovering Qualitative Methods: Field Research, Interviews and Analysis*. Los Angeles: Roxbury Publishing Company.

Williams, D. (2004). Chinatown Is a Hard Sell in Italy. *Washington Post*, March 1.

Zukin, S. (1996). Space and Symbols in an Age of Decline. In King, A. D. (ed), *Re-Presenting the City: Ethnicity, Capital and Culture in the Twenty-First Century Metropolis*. London: Macmillan, 43–59.

Part 2

Contestations and rights

Public and civic

4 Civic landscapes of post-socialist cities

Urban movements and the recovery of public spaces

Mariusz Czepczyński

Public spaces and civic landscapes

In contemporary geographical landscape studies, *cultural landscape* refers to the 'ensemble of material and social practices and their symbolic representation' (Zukin 1993, 16). Cultural landscape becomes a social product, which embodies representations of powers as well as practices and their interpretations. Urban scenery grows to be not only sets of buildings, spaces and places, but also 'expressions of cultural values, social behaviour, and individual actions worked upon particular locations over a span of time' (Meining 1979, 6). Cultural landscape is a palimpsest, representing and reconstructing the relationships of powers and history through the system of signs, written on many layers, including aesthetic, political, ethical, economic, infrastructural, legal and many others (Cosgrove and Daniels 2004; Black 2003). Representations through landscapes are therefore central to the process by which meaning of space is produced (Hall 2002). The landscape idea represents a way of seeing in which people have 'represented to themselves and to others the world about them and their relationship with it, and through which they have commented on social relations' (Cosgrove 1984, 1).

Each social/political/cultural formation creates its own cultural landscape, always representing the shared civil or public values. *Civics* refers usually to citizens and their obligations and rights. The society based on civic values becomes a civil society, founded on collective action around shared interests, purposes and values. In theory, its institutional forms are distinct from those of the state, family and market, though in practice the boundaries between state, civil society, family and market are often complex, vague and negotiated. Civil society commonly embraces a diversity of spaces, actors and institutions. Landscapes facilitated by civic values, functions and forms might be called a civic landscape. Henri Lefebvre (1991) suggests that each society 'secretes' its own space, through values coded in cultural landscape features. Public space, whether expressed as such or not, is directly 'secreted' by the city as a social system into its very physical fabric. Understanding the ways in which this osmosis between societal and physical spaces is enacted in public space is very important for our understanding of the city and society at large (Burte 2003).

Civic space is a primary public good, which is a base of a well-functioning civic society. Civic spaces represent, constitute, and enhance the daily lives of citizens. As such, they hold an important role in constructing communities' social values regarding sustainable use of shared resources (Abrahamson 2008). Civic spaces are an extension of the community. When they work well, they serve as a stage for our public lives. If they function in their true civic role, they can be the settings where celebrations are held, where social and economic exchanges take place, where friends run into each other and where cultures mix. When cities have thriving civic spaces, residents have a strong sense of community; conversely, when such spaces are lacking, people may feel less connected to each other. Great civic spaces are recognised and valued in their cities as places with their own special flavour that relate to and nurture the larger community and bring the public together (Project for Public Places 2014). Everyday uses of common places are at the very heart of the creation of social linkages and relationships. A place shared together represents universal values and carries collective responsibilities. Traditional market squares, streets and commons make urban life possible, as a place of negotiations and interactions.

Civil landscape can be seen as a substantial element of a public good; a non-excludable and non-rivalrous good in a sense that individuals cannot be effectively excluded from its use and where the use by one individual does not reduce its availability to others. Paul A. Samuelson (1954) defined a public good, or as he called it a "collective consumption good", as a good which all enjoy in common in the sense that each individual's consumption of such a good leads to no subtractions from any other individual's consumption of that good. A good which is rivalrous but non-excludable is sometimes called a common-pool resource. Public space and civic landscape become a quintessentially urban common-pool resource. All the places where the general public can exercise their civic rights and obligations can be called civic, while the respective urban setting, enriched with connotations and forms, becomes civic landscape.

Civic landscape is a very temporal and changeable structure. Its major functions always follow current needs and expectations of a given group of users and stakeholders. Mutability and fluctuation of meanings are always reflected in the instability of its social and cultural functions. The recent civil landscapes discourse shows significant transformation in its course. Former class tensions of the industrial cities have been replaced by place tensions of the post-industrial urban regions (Lussault 2009). The place-based tensions and expectations create a discreet network of social and economic relations, which facilitate urban structure and urban life. The class/place debate has been exercised in many squares, streets and parks for decades around the Western world, but the last 10 years involved intensification of the process, especially in Central and Eastern European post-socialist cities.

Post-socialist erosion of the civic landscape

Social and spatial transformations often result from the path of dependent historical conditions. The process is very clearly exemplified in post-socialist Europe, where communist experience of public place and civic rights had influenced public landscape and activities for more than four decades. Socialist civic landscape policy was an important tool for empowering the communist rulers. The forms, functions and meanings of landscape were designed to unmistakeably communicate the relationship of powers. In most of the Eastern Bloc countries the words 'civil' or 'civic' became merely seldom-used adjectives of the state institution, fully controlled by the Party regime. In some countries *civic* turned out to be a synonym of anti-civic function, like in communist Poland, where the Civic Militia (Milicja Obywatelska) was the main apparatus of anti-civic repression and terror until 1990. Landscapes called 'civic' were mostly specifically designed and carefully arranged public meeting places to gather thousands of seldom-volunteer employees, students and school pupils (Czepczyński 2008).

The over-exploitation and habitual misuse of the terms 'civic', 'public' or 'social' resulted in certain indifference to anything 'civic'. The devaluation of the 'civic' had been seen in every socialist country, in practically every aspect of public life. *Civic* became meaningless or, sometimes, even negatively associated with the hypocrisy and bureaucratic power of the Communist Party. Elongated and compulsory exercise of the 'civic' communist landscapes made the later socialist societies quite reluctant towards any actions aimed at the 'common' good or space. *Civic* or *public* were often perceived as something managed, organised and facilitated by and for the Communist Party. This unambiguous discrediting of the 'civic' concept and socialist variation of 'civic landscapes' resulted in resistance to participation in many civic actions in the coming years. Central public places, usually designed for communist rituals, rallies and parades, became the sites of resistance and real civic disobedience. The Stalin Avenue in Berlin was filled by thousands of protesters in June 1953, like Stalin Square in Budapest in October 1956, Wenceslas Square in Prague in 1968 or a square outside the Lenin shipyard in Gdańsk in 1970 and 1980. Public places rose to be particularly important, or even crucial, during the 'autumn of the people' in 1989. In many cases the collapse of communist regimes was initiated, facilitated or just happened in public areas that were turned into prototypical civic places, including Nikolaikirchhof in Leipzig, Wenceslas Square in Prague, and, above all, the Palace Square in Bucharest.

After the collapse of socialism, cities faced vast legal, economic and social conversions. Changes have been accelerated by the expansion of free markets and flows of capital, the reintroduction of land rent, and privatisation, as well as the appearance of new actors on the scene of public space, including local governments, free media, private owners and investors, plus citizens and non-governmental organisations (NGOs). Post-socialist revolutions of the early 1990s were characterised by a fairly spontaneous understanding of freedom on both personal and institutional levels. After more than 40 years of oppression and restraint, the control mechanisms almost disappeared. Radical changes in urbanised landscape administration resulted in spatial confusion and place anarchy. The politically, socially and, above all, economically liberated citizens enjoyed the rights of private ownership, to an extent seldom met in Western European countries. The feast of neoliberal urban life was rarely associated with the urge for civic responsibility, compassion or willingness to compromise. The common symbols of public space are increasingly derived from the nexus of aesthetic display and commercial culture. In many Central European cities, similarly to some cities in Western Europe and the US, the projects of 'public space' include growing numbers of coffee bars, 'Disneyfied' streets, and large interior shopping complexes that attempt to provide an overarching spatial metaphor for social identity (Zukin 1998).

Public space is rapidly disappearing, either through the processes of globalisation and privatisation or through new forms of social control, such as policing and video surveillance. Even main market squares, the central, civic spaces of most of the historic Central European cities, are being redesigned, and regimented in ways that restrict their traditional social and political uses. In this discussion the concerns of the users are contrasted with the intentions of the designers and government officials, in order to highlight how the conflict between representational and use value of public space is worked out in a specific context (Low 2005). Renegotiations of civic rights in public spaces and abandonment of civic responsibilities mirrored the ongoing economic and cultural transformation of the Central European societies. New exclusions and hierarchies are constructed in space and landscape, by means of property rights and transportation systems. Dominant economic and political institutions carve their imprint on the urban landscape by producing what Lefebvre (1991) calls 'abstract space'. This space is delineated and defined by capital investment and prestigious public and private projects that have brutalised and dominated the city (Zukin 1993). New quasi-civic landscapes of Central and Eastern European cities had been designed to meet the needs and expectations of one emerging social group: the up-and-coming middle class. 'Glamorous and beautiful' landscape symbols dominated urban squares, malls, fountains, roads, car parks and pedestrian streets. Squares became shopping malls, lawns were turned into paid car parks, and urban meadows and small gardens were transformed into gated residential 'communities'. Public places radically shrank in the 1990s in practically every city of the region. The citizens became insatiable consumers of every possible good: not only food, clothes, cars and houses but places and spaces as well. Public or civic good was turned into a consumable good, with its price and economic value.

Neoliberal policies carried many threats to civic societies and spaces. Semi-public, secure, clean and functional places had been constructed. Those places often eliminate many forms of civic use, but create only a public place simulacrum, which from a distance seems to be public, but often it is a no-person place or autonomic space. While the high streets are dying, both functionally and aesthetically, drained functionally by the malls, the cities shrink not only spatially and quantitatively but also qualitatively, becoming large suburbs with empty or dangerous centre-less centres. The concentrations of exclusive malls, middle-class gated districts and CCTV-controlled streets and squares turn the other streets and districts into a no-go area. Many urban planners, municipal managers and local leaders seceded the process of place-making to the private sector. It was a typical mantra of the 1990s that the developers are those who (shall) develop the city. Urban policy had often been facilitated by practically omnipotent developers. The economic vitality, development boom and practically unrestricted planning resulted in grotesque urban landscapes, often pretentious and sometimes play-acting at 'civic'. Many post-communist countries, lost in liminal times and spaces, face the withdrawal of the public and civic sphere. A broken balance between public and private space is the result of both weak public governance and fragmented societies. Lost confidence in the public city and landscape can be seen as a specific post-communist overreaction, and citizenship as a newly gained privilege, and as such it includes and excludes politically, socially and spatially. Strict implementation of neoliberal economic and spatial policies makes public spaces less public, more commercial and placeless, while the public demand for public spaces had decreased during the 1990s.

Spaces of political manifestations or reclaiming civic landscapes

Since the late 1960s, in many Western countries regimes of public or civic space in all cities have become more democratic, inclusive and tolerant. A similar process, but often a faster one, has occurred in Central European cities since 1989. Political demonstrations, performances and festivals have made public spaces more dynamic, more attractive and more open to different social groups. But public spaces have also attracted people who were once limited to narrow areas of the city, including drug users, dealers in illegal or illicit goods, and the homeless. For the past 25 years, these combined conditions of festivity and dereliction have led first to a devalorisation and then, in some cities, to a revalorisation of public space, primarily at the city's centre. 'Valorisation' of urban space occurs in several senses: financial, in terms of property values; moral, in the sense of social values; and visual, in aesthetic values suggesting 'good' or 'bad', 'dangerous' or 'safe', evoked by relationships between public spaces' users, uses, and designs. Financial, moral and visual values are always connected. During the 2000s, in many post-socialist cities, the desire to re-establish an aura of social coherence in central urban spaces – to make them more inspiring, and more profitable – encouraged various movements to re-aestheticise the centre and to implement civic values in the urban space (Zukin 1998). Since the early 2000s a certain erosion of neoliberal public space practices had become visible in many Central European cities. The change was facilitated by both systematic enrichment of the urban societies and, probably even more, by the growth of a new class of conscious urbanites. The process was accelerated by the accession of the former communist countries into the European Union in 2004 and 2007. New aspirations and civic needs of the new class met new modes of operations, while practically unlimited flows of ideas and people between the 'Western' and 'Eastern' part of Europe raised the expectations of urban societies.

Public spaces play an important role in civic and political discourses, while spontaneous and bottom-up movements have often facelifted the process in the public sphere. Democratic expressions of civic practices include public strikes, protest marches or political gatherings, often organised in front of the new centres of powers. Parliaments are now being recognised as the main decision-making institutions, so the protests are organised usually on squares in front of the buildings, like the 'white town' – a set of tents of protesting nurses in Warsaw – or other famous protests in Budapest and other Hungarian cities in the autumn of 2006. Public places like Budapest's Kossuth and Szabadság squares were the scenes of mass gatherings in September 2006 to protest against the government. Those places or landscapes are especially important agoras in the turbulent times of political debates and transformations, when manifestations and gatherings are a very important ritual and way of negotiation. The significance of new public civic landscapes of discourse has also been recognised by politicians, who decided to completely close Szabadság Square in Budapest to prevent meetings, while hundreds of people camped in a tent city on the nearby Kossuth Square to prevent the police forces overtaking 'their' spaces of manifestations (Czepczyński 2008).

Constructed during the communist era, centrally located anti-civic places of manifestation swiftly regained their civic functions in many cities around the region. Social protests, workers' conventions, gay-pride marches and pro- and anti-government rallies occupied major plazas and boulevards. Public demonstrations have been practised on the historical squares and streets, while probably the most famous is Kiev's Maidan Nezalezhnosti (Independence Square), formerly known as October Revolution Square. As the central Kiev square, the Maidan has been the centre of public political activity following the end of the Soviet era. In the autumn of 1990, students' protests and hunger strikes in the Maidan resulted in the resignation of the Ukrainian prime minister. During the Orange Revolution in late 2004, Maidan Nezalezhnosti received global media coverage as hundreds of thousands of protesters gathered in the square and pitched tents for several weeks to protest against electoral fraud (*Kiev Encyclopaedia* 2008). During the winter of 2013/2014 Maidan, better known as Euromaidan (Euro Square), was a focal point for civil unrest in Ukraine, which began on the night of 21 November 2013 with public protests demanding closer European integration.

The protests reached a climax during mid-February 2014. On 18 February, the worst clashes of Euromaidan broke out after the parliament did not accede to demands that the Constitution of Ukraine be rolled back to its pre-2004 form, which would lessen presidential power. Police and protesters fired guns, with both live and rubber ammunition. The fights continued through the following days, in which the vast majority of injuries took place. On the night of 21 February, Maidan vowed to go into armed conflict if President Yanukovych did not resign by 10 a.m. Subsequently, the riot police retreated and Yanukovych and many other high government officials fled the country (Calamur 2014). The next day, the parliament removed Yanukovych from office and replaced the government with a pro-European one. In the aftermath, the Crimean crisis began amid pro-Russian unrest. For a year, Maidan (the square) became more of a symbol of democratic and independence movements. The physical square remains a powerful civic landscape icon of Ukrainian civic society.

Practically all civic spatial manifestations had a local, urban or national character. Only one major movement gathered thousands in almost every major city of the region: from Gdańsk to Sofia, and from Ljubljana to Tallinn. The Anti-Counterfeiting Trade Agreement (Acta) brought a new dimension of civil landscapes, especially of often rather passive youngsters. The agreement aims to establish an international legal framework for targeting counterfeit goods, generic medicines and copyright infringement on the internet. After Poland's announcement on 19 January 2012 that it would sign the treaty on 26 January, a number of Polish government websites were shut down by denial-of-service attacks that started on 21 January. Over a thousand people protested in front of the European Parliament office in Warsaw on 24 January. On 25 January, at least 15,000 demonstrated in Kraków, 5,000 in Wrocław, with considerable protests in cities across the country. Similar action had been undertaken in Slovenia, where 3,000 Slovenians subsequently protested at Congress Square in Ljubljana and around 300 in Maribor on 4 February 2012. Protests attracted thousands on the main squares in number of Central European cities, including Bulgaria, where up to 8,000 protested in Sofia. In Croatia, protests were held in Zagreb, Split and Rijeka, with demonstrators, some masked, carrying banners reading 'Stop internet censorship'. In Prague, in the Czech Republic, about 1,500 people marched against Acta. Some waved black pirate flags with white skulls and crossed bones, and others wore white masks based on Guy Fawkes. In Bratislava, hundreds of young Slovaks rallied, many also wearing Guy Fawkes masks. About 1,000 people demonstrated in Budapest (Arthur 2012).

Better urban places or towards participative and discursive place-making

Civic spatial awaking of post-socialist European societies has been manifested by numerous, mostly local initiatives. Residents, activists and citizens started to take responsibility for their urban places. Their actions were often initiated by a single person – an artist or a young researcher – or a few annoyed neighbours who were gathered to protest against one arrogant use of local power too many, or for one lawn to be saved or one place to be reclaimed.

Engaged public art or urban movements were often bottom-up initiatives, which aimed for more than purely the aesthetic or economic values of landscapes. The transformation is a part of the broader process of regeneration of civic responsibility, visualised in spatial actions.

Civic actions against privatised or neglected public spaces frequently start with protection of green areas, as they are traditionally the easiest targets of market pressure. The uncertain status of the remaining open spaces has led to a feeling of distrust from the inhabitants against the local authorities and the new owners. The process is often exercised and visualised in urban landscapes. In 2007 the anonymous Grupa Pewnych Osób (Group of Some Persons) illegally planted grass on centrally located but neglected squares. The deed was organised at night, well documented and promoted by the internet and local media in Poland's third largest city of Łódź. Later they took action to 'civilise our spaces, to be more friendly and to make us feel good' (Wodecka 2008, 12) by monitoring and listing neglected and derelict places, and pressing local media and local government, often with positive responses (Grupa Pewnych Osób 2008).

Similar movements appeared in many cities around the region, including Sofia, as highlighted by Rode (2007). In reaction to plans designed to develop vast open spaces near the Mladost housing estate, a civic initiative was established in 2001, which aimed to protect the common interests of the inhabitants. In 2002, the group was registered as an NGO. Within a short period, similar initiatives were founded all over Sofia; their number grew to 40 by the beginning of 2006. 'Green Sofia' was established – a civil movement for the protection of open and green spaces in the city. In 2005 the Network of Associations of Citizens of Sofia was registered. With the civic organisation of resistance against the governmental regulation of public open space, a bottom-up process has been started, which has the potential to change the governing technique at the local level in Sofia. The communal elections at the end of 2005 brought a change in the city municipality and in the attitude towards open space and towards the participation of civil society. The issue of Green Sofia has become a serious theme of administrative work, and representatives of the political administrative system are establishing regular contacts and meetings and cooperation with the civic movement. The case of Mladost exemplifies the direct spatial and social effects of the restitution process on open spaces. This shows the close interconnectivity of the mode of regulation of open space with the emergence of civil society structures (Rode 2007).

Another example of growing public-space responsibility comes from the small seaside town of Sopot in Poland. The Sopot Development Initiative has been active in the city since December 2008. It was formed by a group of a dozen or so Sopot citizens who believed that it was important for locals to have the opportunity to help decide about city development projects. One of the initiative's first undertakings was Grodowy Park, the town park. A public consultation exercise was conducted on an urban planning study of the area where the park is located. Citizens made it clear they wanted the park area to remain green and that it should not be built on. The authorities assured residents that the decisions taken at the consultation would be binding, but it transpired that the land had been divided in plots and prepared for sale. The results of the consultation with citizens were not considered. In October 2009, the Sopot Development Initiative launched a campaign 'Democracy is more than elections', aimed at demonstrating to the citizens that they had the right to decide about what happens in their city, and not only through voting for their representatives during local elections. Its main objective was to get the city to implement measures enabling co-decision-making by citizens about issues of community concern. In 2010, the initiative raised the need for the establishment of participatory budgeting, and the Sopot city council passed a resolution on enforcing such budgeting on 6 May 2010. This was the first time a city in Poland voted to implement participatory budgeting, and

Sopot has become an example of good practice for other cities (Prykowski 2011). In 2015 more than 70 municipalities around Poland, including all the largest cities, have implemented certain types of participative budgeting. Most of the proposed projects are related to leisure activities, including construction or revitalisation of playgrounds, paths, neighbourhood parks and sport fields (*Ponad 70. . . 2015*).

In spring 2008 the Polish Citizens' Forum initiated a nationwide discussion on the quality of public and civic places in Poland (Radwański 2008). Its slogan, as promoted by one of the largest national newspapers, *Gazeta Wyborcza*, was 'Give us back our cities/parks/streets/playgrounds/windows'. This reflected newly grown civic distress over the privatisation and commercialisation of public spaces. The newspaper, together with the forum, organised a nationwide photographic competition to show examples of the occupation and reclamation of public spaces. The series of articles brought forth many aspects of civic rights and broadened the spectrum of interests. The visible external space, which is also private but visually accessible for everybody, also has a common good and should be controlled. There is clearly a growing feeling that limits of personal freedom should be restricted by civic responsibility (Pawełek 2008).

The socially engaged public became a part of spatial political discourses, mastered by renowned Czech artist David Černý, who gained notoriety in 1991 by painting a Soviet tank pink to serve as a war memorial in central Prague. His most recent project includes a 10-metre statue on a pontoon boat: in October 2013 the controversial artist known for his anti-communist stance sent a very clear message to the republic's president ahead of parliamentary elections by installing a giant purple hand with a raised middle finger on Prague's main river, the Vltava (RT.com 2013).

Another example comes from Warsaw, where the installation of the Tęcza (rainbow) has led to nationwide discussion and protest. The artistic construction in the form of a giant, nine-metre-tall rainbow made of artificial flowers, designed by Polish artist Julita Wójcik, has been located on the Saviour Square (Plac Zbawiciela) since summer 2012. It has been vandalised several times, generating significant media coverage in the Polish media, usually in the context of lesbian, gay, bisexual and transgender (LGBT) rights in Poland. What was intended as a work of public art – without an overt political message beyond the need for inclusiveness, according to the artist behind it – has instead become part of a cultural war over homosexuality that has been brewing in one of Europe's most Catholic countries. Originally covered with some 16,000 artificial flowers, the rainbow quickly turned into a giant Rorschach test for residents. To some, the rainbow is just a rainbow, all about 'the happiness in life'. To Włodzimierz Paszyński, the deputy mayor of Warsaw, 'It's a sign of unity; it evokes warm feelings.' Ms. Wójcik, 42, said: 'The rainbow is not a pro- or anti-gay declaration. It's about tolerance, diversity, openness' (Kozlowska 2013). Since its installation the rainbow has been set on fire four times and come under attack from right-wing politicians and websites. As the rainbow symbol is also associated with the LGBT movement, locating the Tęcza in the Saviour Square in Warsaw has proved controversial, and for some right-wing and conservative politicians the 'faggot rainbow' as a 'symbol of deviancy' 'had hurt the feelings of believers' (Przybyszewski 2013).

Conclusions

Civic engagement is based on the encouragement of the general public to become involved in the political process. It is the community coming together to be a collective source of change. Civic engagement is about the right of the people to define the public good, determine the policies by which they will seek the good, and reform or replace institutions that do not serve that good. Civic engagement can also be summarised as a means of working together to make a difference in the civil life of our communities and developing a combination of skills, knowledge, values and motivation in order to make that difference. It means promoting a quality of life in a community through both political and non-political processes (Ehrlich, 2010). Civic engagement always comes from civic responsibility, which grows slowly but steadily, transforming urban spaces into the civil landscapes.

Post-socialist societies have been evolving since the fall of the Wall. From drained subjects and members of the masses during the communist era, citizens were turned into frenetic consumers (see Needham 2003). In this post-traumatic post-polis new citizens of those new post-totalitarian cities increasingly felt like merely 'guests and aliens' (Sassen 2000). Since the early 2000s new social, cultural and political processes have been visible in many of the cities of the region. The 'guests' began to behave like domesticated and responsible hosts, and the 'aliens' started to know and respect each other. The vibrant recovery of civic rights and places still requires education and negotiation by conscious and responsible citizens. More and more of us seem to notice that the value and quality of landscape, and especially its civic and public aspects, directly influence our quality of everyday life. This important change in the character of life in present-day public spaces underlines the importance of creating high-quality spaces, which invites the citizens to come and to participate.

References

Abrahamson, W. H. (2008). *New Civic Landscapes: Manifesting Cultural Sustainability and Civic Myth in Urban Public Spaces*. Cambridge, MA: MIT Press.

Arthur, C. (2012). Acta Criticised after Thousands Protest in Europe. *The Guardian*, www.theguardian.com/technology/2012/feb/13/acta-protests-europe [accessed November 24, 2014].

Black, I. S. (2003). (Re)reading Architectural Landscapes. In Robertson, I. and Richards, P. (eds), *Studying Cultural Landscapes*. London: Arnold.

Burte, H. (2003). *The Space of Challenge: Reflections Upon The Relationship Between Public Space And Social Conflict In Contemporary Mumbai* [webpage] http://urban.cccb.org/urbanLibrary/htmlDbDocs/A014-C.html [accessed November 20, 2014].

Calamur, K. (2014). 4 Things To Know About What's Happening in Ukraine. NPR. www.npr.org/blogs/parallels/2014/02/19/279673384/four-things-to-know-about-whats-happening-in-ukraine [accessed November 20, 2014].

Centre for Civil Society (2004). What is Civil Society? London School of Economics. http://www.lse.ac.uk/collections/CCS/what_is_civil_society.htm [accessed November 6, 2014].

Cosgrove, D. E. (1984). *Social Formation and Symbolic Landscape*. London and Sydney: Croon Helm.

Cosgrove, D. E. and Daniels, S. (eds) (2004). *The Iconography of Landscape: Essays on the Symbolic Representation, Design and Use of Past Environments*. Cambridge: Cambridge University Press.

Czepczyński, M. (2008). *Cultural Landscape of Post-Socialist Cities: Representation of Powers and Needs*. Aldershot: Ashgate.

Ehrlich, T. (ed) (2010). *Civic Responsibility and Higher Education*. Phoenix: Oryx Press.

Grupa Pewnych Osób (2008). http://gpo.blox.pl/html/1310721,262146,21.html?0 [accessed November 10, 2014]

Hall, S. (2002). The Work of Representation. In Hall, S. (ed), *Representation: Cultural Representation and Signifying Practices*. London, Thousand Oaks, CA, and New Delhi: Sage.

Kiev Encyclopaedia (2008). Maidan Nezalezhnosti. http://wek.kiev.ua/uk [accessed November 6, 2014].

Kozlowska, H. (2013). Rainbow Becomes a Prism to View Gay Rights. *New York Times*, March 21. www.nytimes.com/2013/03/22/world/europe/in-warsaw-rainbow-sculpture-draws-attacks.html?_r=0 [accessed November 10, 2014]

Lefebvre, H. (1991). *The Production of Space*. Oxford: Blackwell.

Low, S. (2005). *Transformaciones del espacio público en la ciudad latinoamericana: cambios espaciales y prácticas sociales*. www.bifurcaciones.cl/005/Low.htm [accessed September 5, 2014].

Lussault, M. (2009). *De la lutte des classes à la lutte des places*. Paris: Grasset.

Meining, D. W. (1979). Introduction. In Meining, D. W. (ed) *Interpretation of Ordinary Landscapes: Geographical Essays*. Oxford: Oxford University Press.

Needham, C. (2003). *Citizen-Consumers: New Labour's Marketplace Democracy*. London: Catalyst.

Pawełek, K. (2008). Oddajcie nam nasze okna. *Gazeta Wyborcza* 134.5744 23.

Ponad 70 samorządów z budżetami na 2015 rok (2015). Budżety Obywatelskie. http://budzetyobywatelskie.pl/info/237/ponad-70-samorzadow-z-budzetami-na-2015-rok/ [accessed November 6, 2014].

Project for Public Places (2014). What is a Great Civic Space? www.pps.org/reference/benefits_public_spaces [accessed November 6, 2014].

Prykowski, L. (2011). Public Consultations and Participatory Budgeting in Local Policy-Making in Poland. In Forbring, J. (ed), *Learning for Local Democracy: A Study of Local Citizen Participation in Europe – Central and Eastern European*. Vienna: Network, 89–107.

Przybyszewski, R. (2013). „Pedalska tęcza". Kownacki podtrzymuje swoje słowa. November 15. *TVP Regionalna Bydgoszcz*. http://regionalna.tvp.pl/13020899/pedalska-tecza-kownacki-podtrzymuje-swoje-slowa [accessed November 20, 2014].

Radwański, W. (2008). Kto ukradł nam boisko. *Gazeta Wyborcza* 101.5711 18.

Rode, P. (2007). *Contested Space in Post-Socialist Sofia – Reading Negative Space as Urban Landscape*. www.rali.boku.ac.at/9797.html [accessed November 23, 2014].

RT.com (2013). Provocative art: Rebel sculptor gives Czech president the finger ahead of polls. October 22. http://rt.com/news/czech-floating-finger-elections-548/ [accessed November 16, 2014]

Samuelson, P. A. (1954). The Pure Theory of Public Expenditure. *Review of Economics and Statistics*, 36(4): 387–389.

Sassen, S. (2000). *Guests and Aliens*. New York, NY: The New Press.

Wodecka, D. (2008). Nie oddam mojego miasta. *Gazeta Wyborcza* 215.5825 12.

Zukin, S. (1993). *Landscapes of Power: From Detroit to Disney World*. Berkeley and Los Angeles, CA, and London: University of California Press.

Zukin, S. (1998). Politics and Aesthetics of Public Space: The 'American' Model. In *Ciutat real, ciutat ideal. Significat i funció a l'espai urbà modern*. Urbanitats 7. Barcelona: Centre of Contemporary Culture of Barcelona.

5 Public space, memory and protest during post-socialist transformation

The emergence of Piaţa Universităţii (University Square), Bucharest, as a space of protest

*Craig Young, Duncan Light and
Daniela Dumbrăveanu*

Introduction

From the 'Orange' and 'Ukrainian' revolutions in Maidan Square, Kiev, in 2009 and 2014 (Beissinger 2011; Way 2014), through marches supporting tolerance and equality for lesbians, gay men and other sexual dissidents in Polish cities (Binnie and Klesse 2013; Binnie 2014), to residents of St. Petersburg, Russia, exercising their 'right to the city' to protest against inappropriate urban development (Dixon 2010), and citizens of Belgrade/Beograd resisting the Milosevic regime in 1996–97 (Jansen 2001), public space has become a vital arena for various forms of protest in post-socialist cities across the former Eastern Europe and Soviet Union. These apparently 'public' spaces within cities have come to play a central role in complex processes of developing civil society and democracy in the context of the post-authoritarian, or even semi-authoritarian, socio-political systems which emerged after 1989–91. However, while much research has worked to unpick the role of urban public space in various movements espousing a 'right to the city' (Lefebvre 1968, 1996; Mitchell 2003; Harvey, 2008) and international social movements such as 'Occupy' (Kilibarda 2012; theme issue of *Journal of Critical Globalisation Studies* 2012, 5; Uitermark and Nicholls 2012) in a Western, capitalist context, relatively little is known about how public space has emerged as a site of protest in a post-socialist setting, even though some of those societies could now be considered capitalist and even 'Western'. As Dixon (2010) suggests, such struggles are a part of post-socialist societies' efforts to create new polities and identities.

This chapter therefore presents a case study of the historical development of protest in Piaţa Universităţii (University Square), in the Romanian capital Bucharest, in order to explore the role that urban public space plays in society and politics in a post-socialist context. From its origins as a public space in the late nineteenth and early twentieth century under the Romanian monarchy, the square underwent various changes during the communist period and then again after the Romanian 'revolution' of 1989, as subsequent political regimes sought to shape the meanings attached to this space and as it became associated with major historical events linked to protests against both communist and post-socialist regimes. And since 1989 these processes in this one public space have also been influenced to different degrees by larger-scale processes of a 'return to Europe' and European Union (EU) accession, globalisation, the global economic crisis in 2008–09 and the growth of international protest movements. In this context, the chapter addresses some key questions about post-socialist public space, including: what factors shaped particular public spaces as spaces of protest during communism, the fall of state-socialism and then post-socialism; how did the particular circumstances of the fall of communism shape the nature of public space as a space for protest under post-socialism; and what does this say about the role of public space in post-socialist civil society and democracy?

As work on post-socialist urban spaces has explored, there are a range of questions to be addressed about the inter-relationship between civil society, protest and democracy as expressed and performed in public space (Way 2014), to which could be added the idea that the nature of such events in public space can also say a lot about the nature of post-socialist governing regimes. Writing as early in the process of post-socialist transformation as 1993, Bernhard (1993, 326) concluded that 'the successful democratization of Soviet-type regimes will include the reconstitution of a civil society as a means to curtail state autonomy and as a basis for a new system of interest representation', and that this will vary between differently configured civil societies.

Mitchell, following Lefebvre (1968), argues that the playing out of the relationship between civil society and democracy is inherently spatial, as publicly expressed concerns over the 'right to the city' are often about power struggles between those seeking to impose (or resist) order and control over (public) space, and 'that order must be explicitly geographic: it centres on the control of the streets and the question of just *who* has *the right to the city*' (Mitchell 2003, 17). Mitchell and Staeheli (2005, 798) advance this point further by arguing that 'public space is where dissent becomes visible. The question is, then: What are the conditions under which visibility becomes possible?' In their view, publics (and civil society and democracy) are in part constituted in and through public space, and 'The politics of public space, therefore, can shape the nature of politics in public space.' Here, *the politics of public space* refers to how it is controlled, for example by legislation and policing practices, and how this shapes the ways it can be used to express dissent.

However – without wishing to consign everything that happens in public space to a simplistic category of 'resistance' – this politics is also about how the streets and urban public space form both a *specific* terrain and a representational space in which power can be contested (Jansen 2001; Routledge 1997). As Jansen asks about the 1996–97 pro-democracy protests on the streets of Beograd:

> Why did they come about when they did, and why were they concentrated in cities, and especially in the Serbian capital Beograd? How did this specific locale, location and sense of place . . . inform and reflect the character, the dynamics and tactics of the events? What kind of place-specific discursive practice of protest was developed . . .?
>
> (Jansen 2001, 38)

As Uitermark and Nicholls (2012) reveal in their analysis of the international 'Occupy' movement from 2011, its relative success and sustainability in different cities relied heavily on whether Occupy activists could connect with local activist networks and align themselves with their *local* concerns. Urban public spaces are dynamic and how they operate is shaped by local factors in combination with the national and transnational/international. Their ability to sustain their role as loci of protest owes much to their specific accreted discourses, values, meanings and affective registers and how these are produced and reproduced through processes such as memory.

To address these issues the chapter first briefly sketches the historical development of Piața Universității in the period before the establishment of Romania as a communist country (up to 1947) and then during the state-socialist period (1947–89) itself, outlining its role in the urban morphology and socio-political life of the city and the nation. The next section then explores the role of the square in the events that led to the downfall of Romanian communism, the 'revolution' of 1989. These events, and those which quickly followed in the form of the also-violent Mineriadă in 1990, were crucial in shaping how Piața Universității worked, and continues to work, as a space for protesting against regimes. Throughout this account we also highlight how it was not simply the events themselves, but also how they were subsequently memorialised in this space and how they shaped people's memories that make this space significant as a site of post-socialist protest. We then conclude the analysis with a consideration of how Piața Universității, having become associated with protest through these events, has continued to be a site of protest against different post-socialist governments and specific issues, but at the same time is also a public space in which other events are celebrated, suggesting a hybrid space in which many issues and emotions are addressed, not just protest. The chapter concludes by summarising the key characteristics of this urban space and its place in Romania's post-socialist transformation, particularly the relationship between civil society and the state.

The origins of Piața Universității and the square in the communist period

Piața Universității is a major intersection in the centre of Bucharest, but the name is also loosely used to refer to a larger irregular area surrounding the intersection itself (see Figure 5.1). The origins of the square date from the early twentieth century and had little to do with notions of claiming public space for protest, but a lot to do with the state seeking to control public space to project their imaginings of 'the nation'. At this time Romania was a monarchy and had gone through a period from the late nineteenth century of seeking to challenge predominant external perceptions of the country as underdeveloped, even backward, and as peripheral to Europe, both geographically and culturally. Romania had also already gained an image as a rather liminal space, between the West and the East and not clearly belonging to either, but also between the civilised world and the supernatural, fuelled in the West particularly by the popular success of Bram Stoker's *Dracula*. In response, the first Romanian king, Carol I, initiated a process of nation-building which aimed to place Romania as a modern nation-state firmly located within Europe (Boia 2001).

It was in this context that the intersection was created during the early twentieth century in an energetic period of modernisation of the city which, after 1878, had become the capital of independent Romania. The city's leaders were keen to remake Bucharest into a modern European city and, given Romania's historical and cultural allegiance with France, the principal inspiration was Paris. In particular, Haussmann's grand boulevards in Paris were icons of modernity and Romania was eager to imitate them. Thus a major west-east boulevard was completed in 1895 (Giurescu 1976) and named after King Carol I and his wife Elisabeta (see Figure 5.1). In the early twentieth century a north-south boulevard was added, with the intersection between them resembling Haussmann's *grand croisée* in Paris (Celac et al. 2005). To add to the

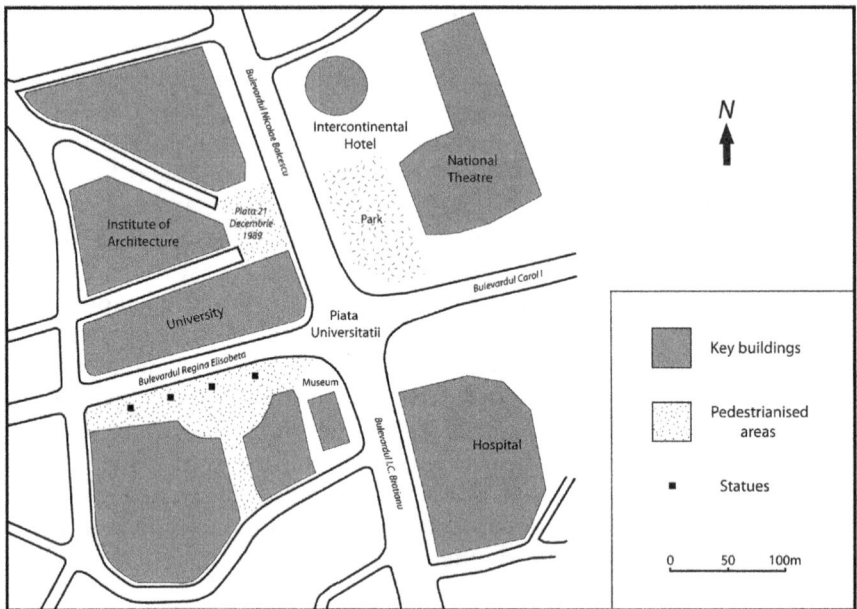

Figure 5.1 Piaţa Universităţii, Bucharest

symbolism, Bucharest's main university building (opened in 1869) stood at the intersection and a number of other grand buildings were constructed around the square.

The intersection was therefore constructed as a statement of modernity and of the national identity which Romania was seeking to cultivate and project, and these efforts were further emphasised by efforts to make it a place of national memory. Four statues of important historical and cultural figures were erected on the south side of the intersection, while a statue of the Liberal politician and nation-builder I. C. Brătianu was placed in the centre in 1903. Brătianu's name was also allocated to the north-south boulevard, while the intersection was named Piaţa Brătianu. This area became the setting for occasional state ceremonies such as an annual military parade attended by King Carol I on the anniversary of his coronation. It was also a popular location for informal political meetings and protests in the period of the monarchy, particularly around the statues on the southern part of the square (Costescu 2005), though this space was not unique in this sense in Bucharest at this time. However, the main location for public gatherings was Piaţa Palatului (Palace Square), half a kilometre away, where the royal palace was located. Although not planned as such, Piaţa Brătianu effectively became the de facto centre of the city (Boia 2001), something that was institutionalised in 1938 when the 'Kilometre 0' monument (the point from which all distances within the country are measured) was erected nearby. Rather than being strongly associated with protest, in this period the area became important as a key part of the social life of the city. It was a place for promenading, particularly on a Sunday, and meeting friends for talking about life, politics and business over a beer or a coffee. Intellectual discussions took place here, as did informal student gatherings, but also cultural events, such as impromptu performances by singer/actress Maria Tănase or actor Constantin Tănăse. These took place particularly in the southern part of the square around the statues, creating a micro-social geography of the square that persists to the present day.

Following the declaration of the Romanian People's Republic in December 1947 a number of changes were made to the intersection with the intention of de-commemorating the monarchy and instead commemorating historical figures considered exemplary revolutionaries by the socialist regime, in order to signal a new narrative for the Romanian nation. Brătianu's statue was removed (although the others were retained) and the two boulevards were renamed (Light et al. 2002). The intersection was initially named Piaţa Bălcescu (after one of the leaders of the 1848 Wallachian revolution). Plans were developed in the 1950s for a form of 'systematisation' of the square intended to create a new public plaza, although building work did not start until the late 1960s (Ioan 2009). The 22-story modernist Intercontinental Hotel opened on the north side of the intersection in 1971, followed by the nearby national theatre in 1973 (which was given a new façade in the 1980s). In the late 1980s the tram lines which ran along the north-south boulevard were removed when the Universitate metro station opened and the intersection was formally renamed Piaţa Universităţii around the same time. Overall the square was partly remodelled along modernist socialist principles and to promote the achievement of the socialist state, such as technological progress as evidenced by the metro.

However, the cluttered and irregular space was of little use to the socialist regime as a venue for public meetings, parades and displays which, instead, took place in larger public spaces in other parts of the city. However, Piaţa Universităţii was still associated with forms of protest in two main ways. First, although not suitable for state-organised protests itself, such events usually involved large-scale gatherings which moved through the city. Piaţa Universităţii was thus often part of such protests in the sense that they started there or paraded through it between larger sites such as stadia on the way to spaces held to be more significant by the regime, notably the nearby large Piaţa Palatului (Palace Square), where the Romanian Communist Party (Partidul Comunist Român, or PCR) had its headquarters. These *mitinguri* were not protests *against* the state, but were organised *by* the socialist state as protests against the wider 'enemies' of communism – the themes adopted for these events included pollution, inequality, unemployment (since in a socialist country such a thing did not exist by presumption), nuclear weapons, respecting territorial 'integrity' and internal affairs. These themes were chosen to represent the superiority of the socialist state over capitalism, and increasingly so the population could show 'support' for the leadership. Events were organised by the *propagandisti*, members of the Propaganda Department within the PCR, and to begin with were voluntarily attended by the population. However, as the population became increasingly disaffected with the regime, organised protests such as these were increasingly held in sports grounds and arenas where crowd control and surveillance was more manageable and larger-scale and people were 'strongly encouraged' to participate, which often meant that they were transported there from, for example, work places. They became less common after 1980, and from 1983 gathering in Piaţa Universităţii was actively discouraged, including by the university authorities, who suggested that groups of students did not associate there. The use of Piaţa Universităţii for gatherings and any form of debate decreased considerably.

However, this did not mean that all forms of relating to the square, and even protest, ceased entirely. In everyday life and mundane activities citizens develop complex relationships with urban landscapes, which may not mirror what regimes intended.

Thus during this time the square consolidated its status in the imagination of Bucharesters in a variety of ways as the symbolic and emotional heart of the city. The presence of the university meant that this was a lively and energetic social space for young people and students. At least in the early years of the communist regime it remained a place for discussion, debate, exchange of ideas and occasional public protests among the Bucharest intelligentsia, not particularly different from in pre-communist times. The part of the square to the north of the university was (and remains) a popular place for friends to meet. A particular landmark in the square was the large 'university clock' dating from the 1920s. Throughout the socialist era this was a popular meeting point among young people: to ask (or be asked) to meet at the clock was a clear request for a date. In various ways the square became embedded in the emotional lives and geographies of Bucharesters as a place associated (sometimes nostalgically) with youth, freedom and opportunity, and not simply projections of the nation or the values of socialism. Indeed, in the early years of the regime it formed a space in which students met and debated socialist principles.

However, again as disenchantment with the regime grew, the square and in particular the areas around it became more associated with protest, but not the kind of open protest in public space which will be discussed below. As the use of public space even for *mitinguri* declined, and even gathering in groups was discouraged, protest took a different form. From the early 1970s, any form of gathering, social or political, was a chance to carefully criticise the regime; not directly, but through the use of humour and jokes which subtly spoke against Romania's President and communist leader Nicolae Ceaușescu and the regime more generally, though always with a watchful eye for Securitate informers. It also became a way of protesting through the careful sharing of news about the failings and excesses of the regimes. Not everyone knew about such things – some people were informed by Radio Free Romania – and sharing stories became a way of resisting the regime. Increasingly, however, such practices could not be undertaken in public space, but locations around Piața Universității became significant for this form of resistance. The university building was briefly a site of meditation meetings, readings and commentaries by a group called Meditatia Transcedentala, who used the exchange of ideas, meditation and oblique references to criticise the regime, eventually leading to their being removed from their jobs and sent into 'production' (factories) or arrested. Their persecution became well known throughout Romania, fuelling further guarded protest, but not in a form which manifested itself by taking to the streets, until the events of December 1989.

Piața Universității as a space of protest and remembrance after 1989

Thus Piața Universității was established originally as an expression in the capital city of Romania's desires to be seen as modern, progressive and European, a set of values which the communist regime tried to supplant by renaming and changing the landscape of the square. In the pre-war and communist periods it was already a site associated with protest, but to a relatively small degree. As Romania's communist regime became increasingly hard-line under the Ceaușescu regime (1965–89), public protest was suppressed. However, Piața Universității was to become a key site within the city associated with dissidence, resisting regimes and protest through events and practices which developed during the violent overthrow of communism in 1989 and the subsequent events of 1990. These events, and the way that they were subsequently memorialised, played a key role in shaping the memories and identity of the square as a terrain and representational space of protest.

On December 21, 1989, Ceaușescu was jeered and heckled as he tried to address a public rally in nearby Piața Palatului (Palace Square). This large square was located in front of the Communist Party Headquarters building and was used for large-scale staged rallies, such as the one the regime had called to try and quell increasing dissent. However, after the meeting restless crowds did not linger but headed towards Piața Universității (which was already seen as the emotional heart of the city) to join others already gathering there. As citizens started to protest, the security forces opened fire. These confrontations continued throughout the day and night and at one point a barricade of cars was built across Bulevardul Bălcescu and set alight. The security services responded with further brutality, so that by the end of the night 49 demonstrators had been killed in Piața Universității and a further 463 wounded (Siani-Davies 2005). The following day, as crowds stormed the Communist Party Headquarters in Piața Palatului Ceaușescu was forced to flee by helicopter (he was later captured and executed). A group calling itself the National Salvation Front (NSF) assumed power in the name of the people. There followed three days of open conflict on the streets of Bucharest, apparently between the army (which had turned to side with the revolution) and forces loyal to Ceaușescu. Piața Universității witnessed little further action during the revolution but remained in the popular imagination as the trigger point, where the first lives were lost in the struggle to overthrow Ceaușescu.

Thus, in early 1990, Piața Universității became an important site of remembrance for the events of December 1989 and those that died fighting the regime. Improvised ephemeral memorials, wooden crosses, flowers and candles (Beck 1993) were placed there by the families of those who had died and other well-wishers. A small, previously unnamed part of Piața Universității located alongside the University and Architecture School, where many people died on the first night of the revolution, was later renamed Piața 21 Decembrie 1989, and here a number of more permanent memorials were placed in the form of small and unobtrusive plaques and crosses with simple inscriptions such as 'For the heroes of the Revolution, 21–22 December 1989' and 'Here they died for freedom, 21–22 December 1989'. Significantly, these were informal and spontaneous acts of remembrance that were initiated by ordinary citizens and non-state organisations rather than by the state. In Bulevardul Bălcescu, which runs past Piața 21 Decembrie 1989, one of a number of much older stone crosses was inscribed by a local painter with the text 'To the heroes of the revolution'. A large wooden cross was also erected here by a group representing those participating in the revolution, with the blessing of the Patriarch of the Romanian Orthodox Church.

What is key about these practices commemorating the revolution in Piața Universității is that they were, and continue to be, undertaken independently of the state authorities. By contrast, 'official' commemoration of the revolution has centred on Piața Revoluției (Revolution Square – the renamed Piața Palatului). Here a monument was erected in front of the former Communist Party building in 1990 and a second, larger memorial was inaugurated in 2005. While the practices of memorialisation in Piața Revoluției are shaped by the state to remember the 'revolution' as a key event overthrowing communism, the commemorations in Piața Universității focus on the individuals who died in the revolution rather than the event itself. These are deathscapes (Maddrell and Sidaway 2010) in which private grief is publicly displayed through smaller, individual and highly personalised forms of memorialisation. Significantly, official and popular commemorations are very different in form, make use of public space in different ways, and have quite separate geographies. Though Piaţa Revoluţiei and Piaţa Universităţii are closely located public spaces in the city, the performances of memory which take place in them and the emotional and affective geographies which adhere to them are significantly different, something which continues to shape how Piaţa Universităţii functions as a public space today.

Moreover, Piaţa Universităţii's distinctiveness was further emphasised by traumatic events which followed the revolution. By early 1990 it was apparent that Romania had not made a decisive break with communism. Instead, it was clear that the NSF, which had taken power on behalf of the people, was dominated by former members of the communist *nomenklatura* whose commitment to reform was unconvincing. Following the NSF's announcement of its intention to stand in the May 1990 elections, students and young people occupied Piaţa Universităţii in a protest camp which quickly grew in size and popularity. The NSF convincingly won the elections (with 67% of the vote) and its leader Ion Iliescu (a former member of the Central Committee of the PCR) was elected President. This provoked further protest by students so that Iliescu resorted to violence. On June 14, 1990, thousands of miners were brought to Bucharest on specially chartered trains and told that Romania's new democracy was under attack from anarchists, deviants and foreign agents camped in Piaţa Universităţii. The miners marched through Bucharest and on reaching the square savagely attacked the protesters and ransacked the university, with the most brutal violence occurring in Piaţa 21 Decembrie. According to government figures seven people died, but the actual total is believed to be in the hundreds. This shocking event – which became known as the Mineriadă (literally 'Miners' Rage') – demonstrated that the post-Ceauşescu regime was as willing as its predecessor to use violence against its citizens.

This led to a further layer of meaning and commemoration developing in Piaţa Universităţii. A diverse range of memorials have been placed in the square to commemorate the young people killed in the Mineriadă. A marble cross in front of the national theatre bears the text 'In memory, June 1990'. Alongside, a monument dating from 1998 resembles a Romanian 'milepost' (Antonovici 2009) declaring the site to be the 'Kilometre Zero' of freedom and democracy in Romania and a 'Zone free of neocommunism'. This both alludes to the nearby Kilometre Zero monument (as the literal centre of the nation-state) and to a slogan from the 1990 protests when students declared the NSF to be 'neocommunists'. The university building in Piaţa 21 Decembrie bears a memorial plaque with the inscription 'Here students and lecturers fought for freedom and civil rights in December 1989 and April–June 1990'. In the centre of Piaţa 21 Decembrie 1989 is a metal cross erected by a local artist, Constantin Popescu. It bears the text 'For the anti-communist heroes' and invites passers-by to place a flower in memory of those who died. The cross is regularly cared for and repainted, apparently by the painter himself. The wall of the Architecture School opposite was extensively graffitied with protest slogans throughout the 1990s (the graffiti were finally cleared in 2001 when the Social Democratic Party, successor to the NSF, was in power). Even today protest graffiti regularly appear in and around the square, some of which link this space to other instances of state repression, such as 'Rangoon 2009', making it to some extent also a site of transnational protest.

Thus, in addition to being a terrain of protest, Piaţa Universităţii is also a highly significant place of memory in Bucharest, a significant representational space (Lefebvre 1991). It is a site associated with state-sponsored violence against the civilian population by the communist regime and a place where a supposedly post-communist government used similar appalling violence against those who questioned its legitimacy. Indeed, it has the status of a 'sacred space' in post-socialist Romania (Beck 1993; Antonovici 2009), and one which reveals much about the relationships between civil society and the state in post-socialist Romania. As noted above, the state on the one hand, and individuals and civil society on the other, commemorate these events in different ways and in different public spaces in the city. The post-Ceauşescu state has always had an ambivalent relationship with Piaţa Universităţii, particularly when the NSF and its successors were in power (1990–96 and 2000–04). Unsurprisingly, the state has not become involved in commemorating these events in Piaţa Universităţii, and state-led attempts at commemoration in Piaţa Revoluţiei are largely ignored (or even actively ridiculed) by most Romanian citizens. The state has made no attempt to reinscribe the meanings of Piaţa Universităţii (apart from removing graffiti) or to intervene with the alternative, personal acts of commemoration. Instead, Piaţa Universităţii – and in particular Piaţa 21 Decembrie 1989 – has become an informal but powerful site of 'countermemory', i.e. unofficial or unauthorised practices of remembrance which may directly challenge official or elite attempts to construct collective memory (Goldberg et al. 2006). It represents an attempt to rebut the efforts of the political elite to shape what is remembered and how (Legg 2005; 2007). Piaţa 21 Decembrie 1989 is a place which reminds ordinary Bucharesters that the deaths of December 1989 did not bring about the desired political change. This tension between official and popular memory, and between different spaces of memory, was further apparent during the twentieth anniversary of the revolution in 2009. Official ceremonies unfolded in Piaţa Revoluţiei, but it was in Piaţa Universităţii that former revolutionaries and Bucharesters gathered to remember the event.

Piaţa Universităţii and protest beyond the revolution and Mineriadă

The events of the 1989 revolution and the 1990 Mineriadă, and the ways in which they were subsequently commemorated and remembered, thus played a significant role in shaping Piaţa Universităţii as a space of protest symbolic of the continued tension between the state and civil society in attempts to develop Romania as a democratic nation-state. And since those events the square has continued to play an important role in the capital both as a terrain of protest and a representational space for attempts to consolidate post-communist Romanian politics and identity.

In terms of Romania's post-socialist political development, Piaţa Universităţii also became a site of broader resistance to the former communist elite who dominated the Social Democratic Party. In 1996, Emil Constantinescu (a professor at Bucharest University and representative of the centre-right opposition coalition) defeated Iliescu in the presidential elections. Following victory, it was to Piaţa 21 Decembrie 1989 that he came to address his jubilant supporters. In July 1997, President Clinton, accompanied by Constantinescu, addressed an enthusiastic crowd of young Romanians in Piaţa Universităţii and acknowledged the sacrifices for freedom that had taken place there. In later parliamentary election campaigns centre-right parties were keen to appropriate the symbolic capital which is attached to the square as a space of opposition to the former communists who dominated political life in the early 1990s. For example, in the run-up to the 2004 elections, the centre-right 'Justice and Truth' party erected a tent in the square and made it the centre of their election campaigning.

More recently, the association of Piaţa Universităţii with opposition to the successors of the PCR was apparent during the presidential elections of November 2014. After the first round of voting, Piaţa Universităţii was the focus of repeated demonstrations and protests when it became apparent that large numbers of Romanians working abroad had been unable to cast their votes (something interpreted by the protesters as an attempt by the Social Democratic Party to manipulate the final result in favour of its own candidate). After the second round of voting, when exit polls predicted that the Social Democrat candidate (widely expected to win) had been defeated by the centre-right candidate Klaus Iohannis (a Transylvanian German) jubilant crowds immediately gathered in Piaţa Universităţii, and it was at this square that Iohannis later came to greet his supporters.

More broadly, since 1990 Piaţa Universităţii has been a site for diverse performances of freedom and resistance that have reinforced its role as the key symbolic space in Bucharest, associated with celebration as well as protest, though the choice of Piaţa Universităţii as a site of celebration is also a tacit rejection of official, state-led attempts to make Piaţa Revoluţiei the symbolic heart of the capital and the nation. Indeed, Antonovici (2009) argues that it is *the* place where Bucharesters feel they can express themselves freely. On some occasions the square is a place for public celebration in a way which recalls the euphoria of the 1989 revolution. For example, when Romania played England in the 2000 European Football Championship, the match was shown live on a giant screen in Piaţa Revoluţiei. Romania won the match and, on the final whistle, people did not linger in Piaţa Revoluţiei but instead headed as one to Piaţa Universităţii, where a large crowd gathered in the square in a joyful celebration.

Piaţa Universităţii was also an important location for Romania's celebrations when it joined the EU on 1 January 2007. For Romania, joining the EU finally represented a decisive break with the communist past and the culmination of difficult political and economic reforms in the post-Ceauşescu era. There was no better place than Piaţa Universităţii to demonstrate that Romania had moved on from the June 1990 Mineriadă and the dominance of the former communist elite in power. Moreover, the square affirmed that those who had died in the Mineriadă had not done so in vain. Several months before accession an 'EU clock' (recalling the original 'university clock') was placed in the middle of the intersection with a digital display which counted down the days and hours to accession. On the night of December 31, 2006, Piaţa Universităţii was the location for the official celebrations of Romania's accession (led by the president). This cramped and irregular space was entirely unsuited to a mass public gathering, so many people (including two of the authors) were unable to get close enough to see anything. A more suitable location would have been Piaţa Constituţiei about a kilometre away, which reportedly has room for half a million people. But this space, immediately in front of Nicolae Ceauşescu's monumental 'House of the People', has entirely the wrong meanings attached to it.

As a place initially associated with opposition to the presence of a government dominated by former communists, Piaţa Universităţii has also become a broader space of protest addressed to governments of all political colours. For example, in January 2012 the square became the site of public protests against the centre-right government and president. They were triggered by the resignation of the popular deputy health minister in protest at the government's attempt to push through partial privatisation of health services. Crowds protested throughout Romania, and in Bucharest several hundred people did the same in Piaţa Universităţii. The following day their numbers had increased significantly, and violent clashes between police and protesters followed (resulting in many injuries on both sides). The Romanian press quickly drew comparisons with June 1990, leading to the proposed health reforms being quickly withdrawn, thus demonstrating the symbolic power of this space drawing on its history of association with opposition during the revolution and Mineriadă. Moreover, the protests continued, but were now directed against austerity, corruption and an unpopular government and president (Ionita 2012). They continued for several weeks (despite freezing temperatures) but, as one commentator noted, failed to attract widespread public support (despite the general unpopularity of the government), so they did not achieve the scale of 'Occupy' movements in other cities (Ionita 2012).

The following year Piaţa Universităţii was the centre in Bucharest of further nationwide protests. A Canadian company proposed developing an opencast gold mine in the small Transylvanian village of Roşia Montană, and the enabling legislation was due to go through parliament in August 2013. However, protesters sought to highlight the environmental damage which they claimed the project would cause (Mercea 2014). The result was nationwide protests throughout Romania which, according to some commentators, were the largest mass protests since the 1989 revolution (Romocea 2013). In Bucharest, crowds of up to 15,000 (mostly young) people gathered in Piaţa Universităţii. Since their protests were initially ignored by the media, the protesters proved highly effective in using social media to promote their cause, which was a notable feature of protests associated with the international 'Occupy' movement (Kilibarda 2012; Lubin 2012). These demonstrations differed from those of 2012 in their exuberant, joyful and non-violent character, which included performances by actors and classical musicians. Nevertheless, there was a strong anti-establishment current underpinning the protest (Tismaneanu 2013a), again linking to the values and meanings now associated with Piaţa Universităţii. The nationwide protests were successful. In December 2013 both houses of the Romanian parliament rejected the opencast mine proposal in what was hailed as a victory for civil society in Romania. One political scientist observed that the protests in Piaţa Universităţii had represented a return of the spirit of protest of June 1990 (Tismaneanu 2013b).

Today the geography of the square is still evolving and this shapes how the space is used for protest. The area to the south around the statues on Bulevardul Regina Elisabeta (see Figure 5.1) has once again developed as an area for cultural performances, cafes and socialising, rather like it was in pre-communist Bucharest. In part this reflects a form of nostalgia for the era of Bucharest as 'Little Paris', which in turn is bound into reimaginations of Romania's post-1989 'return to Europe' and more recently plans to celebrate Bucuresti 555, a series of events to mark the 555th year of the city, itself a means of promoting a new image for the capital internationally. However, the part of the square formed by Piaţa 21 Decembrie 1989 and in front of the national theatre is still firmly associated with remembrance and protest, both in people's minds and various performances. Any organisation seeking to protest does so in this part of the square, and when the media covers protest it always uses this space as a backdrop.

Conclusion

This chapter has analysed the historical development of the characteristics of a notable space for protest in the Romanian capital, Bucharest: Piaţa Universităţii or University Square. This public space exhibits complex dynamic links between the physical development of the square, the events which took place there and how it functions as a space of representation in which politics, identity, civil society, memory and the notion of 'the public' in a Romanian context have been shaped over time. The analysis has shown how the association of the square with protest is a process of long-term historical development. Originally conceived as a space which symbolised the late-nineteenth and early twentieth-century re-invention of Romania as a Western, capitalist, modern nation, it was physically remodelled and associated with a very different vision of the nation under communism. Some early associations with protest and dissent during these periods developed into the square being associated with resistance to both communist and post-socialist regimes though the violent events of the 1989 revolution and 1990 Mineriadă.

This link was strengthened by the ways in which those events were commemorated and remembered in that space, leaving a legacy linking the square to notions of personal sacrifice in the struggle against powerful regimes which informs how the square is used for protests against the state today. Locally-specific factors played a key role in the development of this public space as a space of protest, and the way that those factors were represented and remembered is important for sustaining it as a place of protest. Thus understanding the 'work' of maintaining public space as a space of protest over time is an important part of any analysis of why certain spaces become produced and reproduced through 'place-specific discursive practice of protest' (Jansen 2001, 38). One key factor that emerges here is the importance of how people perform and sustain the memory of protest and sacrifice that gives this public space almost sacred status.

Analysing Piaţa Universităţii has also allowed us to unravel the inter-connections between public space, the state, civil society and democracy in Romania. This public space represents the division in post-communist Romania between state and civil society in which the majority of the population do not see the 1989 revolution as forming a distinct break from the communist past – as the state wishes to portray it – but instead regard the revolution, the Mineriadă and other subsequent events as demonstrating the continuity in power of former communists and a continued divide between the state and civil society. The lack of convergence between civil society and the state is clearly reflected in this geography of representational public space, in which forms of state-led remembrance and memory differ markedly in form and location from those led by individuals and non-state organisations making up civil society. In terms of the issue of control of public space by powerful elites, it demonstrates civil society exerting a 'right to the city' through developing and sustaining a 'countermemory' in and through public space in the face of a powerful elite anxious to promote other, state-led discourses and practices of remembrance. That Piaţa Universităţii continues to be a site of protest against a range of issues in post-communist Romania demonstrates the power of these associations and suggests that Piaţa Universităţii will be a key public space in which the development of the Romanian state, civil society and democracy can be traced for some generations to come.

References

Antonovici, V. (2009). *Piaţa Universităţii – loc memorial? Sfera Politicii*, 17: 94–99.

Beck, S. (1993). The Struggle for Space and the Development of Civil Society in Romania, June 1990. In DeSoto, H. G. and Anderson, D. G. (eds), *The Curtain Rises: Rethinking Culture, Ideology and the State in Eastern Europe*. New Jersey: Humanities Press, 232–265.

Bernhard, M. (1993). Civil Society and Democratic Transition in East Central Europe. *Political Science Quarterly*, 108(2): 307–326.

Beissinger, M. R. (2011). Mechanisms of the Maidan: The Structure of Contingency in the Making of the Orange Revolution. *Mobilization: An International Journal*, 16(1): 25–43.

Binnie, J. (2014). Neoliberalism, Class, Gender and Lesbian, Gay, Bisexual, Transgender and Queer Politics in Poland. *International Journal of Politics, Culture and Society*, 27(2): 241–57.

Binnie, J. and Klesse, C. (2013). 'Like a Bomb in the Gasoline Station': East–West Migration and Transnational Activism around Lesbian, Gay, Bisexual, Transgender and Queer Politics in Poland. *Journal of Ethnic and Migration Studies*, 39(7): 1107–1124.

Boia, L. (2001). *Romania: Borderland of Europe*. London: Reaktion.

Celac, M., Carabela, O. and Marcu-Lapadat, M. (2005). *Bucureşti: arhitectură şi modernitate. Un ghid adnotat*. Bucharest: Simetria.

Costescu, G. (2005) [1944]. *Bucureştii vechiului regat*. Bucharest: Editura Capitel.

Dixon, M. (2010). Gazprom versus the Skyline: Spatial Displacement and Social Contention in St. Petersburg. *International Journal of Urban and Regional Research*, 34(1): 35–54.

Giurescu, C. C. (1976). *History of Bucharest*. Bucharest: The Publishing House for Sports and Tourism.

Goldberg, T., Porat, D. and Schwarz, B. B. (2006). 'Here Started the Rift We See Today.' Student and Textbook Narratives between Official and Counter Memory. *Narrative Inquiry*, 16: 319–347.

Harvey, D. (2008). The Right to the City. *New Left Review*, 53: 23–40.

Ioan, A. (2009). Piaţa Universităţii. *Observator Cultural*, 427. June 12. www.observator cultural.ro/Numarul-427*numberID_798-summary.html [accessed September 20, 2014].

Ionita, S. (2012). Viewpoint: Romania Protests a Warning from the Street. BBC News. January 18. www.bbc.co.uk/news/world-europe-16610093 [accessed October 31, 2014].

Jansen, S. (2001). The Streets of Beograd. Urban Space and Protest Identities in Serbia. *Political Geography*, 20: 35–55.

Kilibarda, K. (2012). Lessons from ⌧Occupy in Canada: Contesting Space, Settler Consciousness and Erasures within the 99%. *Journal of Critical Globalisation Studies*, 5: 24–41.

Lefebvre, H. (1968). *Le droit à la ville*. Paris: Anthopos.

Lefebvre, H. (1991). *The Production of Space*. London: Wiley-Blackwell.

Lefebvre H. (1996). *Writings on Cities*. Cambridge, MA: Blackwell.

Legg, S. (2005). Sites of Counter-Memory: The Refusal to Forget and the Nationalist Struggle in Colonial Delhi. *Historical Geography*, 33: 180–201.

Legg, S. (2007). Reviewing Geographies of Memory/Forgetting. *Environment and Planning A*, 39: 456–466.

Light, D., Nicolae, I. and Suditu, B. (2002). 'Toponymy and the Communist City: Street Names in Bucharest 1947–1965'. *Geojournal*, 56(2): 135–144.

Lubin, J. (2012). The 'Occupy' Movement: Emerging Protest Forms and Contested Urban Spaces. *Berkeley Planning Journal*, 25(1): 184–97.

Maddrell, A., and Sidaway, J. D. (eds) (2010). *Deathscapes: Spaces for Death, Dying, Mourning and Remembering*. Ashgate: Farnham.

Mercea, D. (2014). Towards a Conceptualization of Casual Protest Participation: Parsing a Case from the Save Roşia Montană Campaign. *East European Politics and Societies*, 28(2): 386–41.

Mitchell, D. (2003). *The Right to the City: Social Justice and the Fight for Public Space*. New York, NY: Guilford Press.

Mitchell, D. and Staeheli, L.A. (2005). Permitting Protest: Parsing the Fine Geography of Dissent in America. *International Journal of Urban and Regional Research*, 29(4): 796–813.

Romocea, O. (2013). Who is Roşia Montană? – Or the Dawn of A New Generation. *Huffington Post*. September 13. www.huffingtonpost.co.uk/oana-romocea/romania-rosia-montana_b_3920165.html [accessed October 31, 2014].

Routledge, P. (1997). Space, Mobility and Collective Action: India's Naxalite Movement. *Environment and Planning A*, 29: 2165–2189.

Siani-Davies, P. (2005). *The Romanian Revolution of December 1989*. Ithaca, NY: Cornell University Press.

Tismaneanu, V. (2013a). *Rosia Montana. Piata Universitatii si spiritul revoltei*, 22. September 6. www.revista22.ro/rosia-montana-piata-universitatii-si-spiritul-revoltei-30746.html [accessed October 31, 2014].

Tismaneanu, V. (2013b). *Piata Universitatii, Rosia Montana si puterea celor fara de putere: Renasterea polisului paralel*. www.contributors.ro/politica-doctrine/piata-universitatii-rosia-montana-si-puterea-celor-fara-de-putere-renasterea-polisului-paralel/ [accessed October 31, 2014].

Uitermark, J. and Nicholls, W. (2012). How Local Networks Shape a Global Movement: Comparing Occupy in Amsterdam and Los Angeles. *Social Movement Studies*, 11(3–4): 295–301.

Way, L. (2014). The Maidan and Beyond: Civil Society and Democratisation. *Journal of Democracy*, 25(3): 35–43.

6 Social characteristics of squares as urban spaces

Ulus and Kızılay squares in Ankara

Nuray Bayraktar

Introduction

Urban spaces are meeting, sharing and communication places where all citizens can go without any discrimination and get together. Within this scope, urban spaces are expected to have a unique characteristic that unites all different groups living in the city, and that enables all these groups to communicate and meet with each other. Squares, which are the most efficiently used elements of urban spaces, are defined as places where all social events take place and are designed for the common use of all citizens. Because of these features, squares, which are considered as a significant component of city culture, demonstrate certain differences in various periods of history and in the city of different subcultures.

Ottoman cities did not have any designed squares. The mosque yards were regarded as the most important urban spaces. Squares were open areas, formed by use within the urban tissue, and had an organic shape. These areas were mainly used by men, as a consequence of women's subordinate social roles. A change of the Ottoman social and spatial structure started with the proclamation of the republic, which meant a new modern order of social life. To ensure this order, cities began to be designed in a unique manner. In Republic cities there was mainly a major axis, and there were squares articulated according to this axis. At these squares, various activities took place, which enabled synergy between women and men and participation of women in the modern social life, a model that was widespread and adopted by different social layers. Squares, which were associated with the government structures, were also important for the republic in terms of being ceremonial places. Thus, squares were addressed as an important locus allowing modern social life. However, in those early years of Turkish modernity squares were not designed for pedestrians, especially in the spatial terms defined in the West. But although they were outdoor areas associated mainly with transportation, they still appeared as new places of social experience within cities.

Ankara, as the first city planned after the declaration of the republic, is an important example of this approach. With the belief that the former capital city, Istanbul, represented the Ottoman State, Ankara was selected as the capital city to represent the Turkish Republic. The intention was that Ankara would be a pioneer and a model for other cities in social and spatial terms.

Ankara, capital city of the republic

The history of Ankara starts with that of Anatolia. However, its importance is mainly derived from its status as a capital city. Upon the signing of the Treaty of Lausanne, borders were drawn and Ankara was nominated as the capital city on October 13, 1923. Later, Ankara gained the identity of the capital of the Turkish Republic upon the proclamation of the latter on October 29, 1923 (Tellan 1997, 58).

Ankara was an Anatolian city with a population of around twenty to twenty-five thousand when it was nominated as the capital city. There was neither electricity nor water, and no green areas in the city nor its outskirts. The soil was arid and infertile (Araz 1994, 349). Following its proclamation as the capital, a rapid change started in terms of infrastructure, official buildings and housing. The city began to modernize with comfortable buildings, asphalted streets, planting, parks with pools, green areas and modern public and private vehicles such as buses and automobiles (İleri 1994, 362).

In parallel to this change, public life began to change as well. The ministry buildings, Congress and other official buildings were erected, and the embassies opened new offices, which resulted in a massive influx of civil servants (Sarıoğlu 2001, 74). This group, composed of official and civil bureaucrats coming from Istanbul, constituted the majority of the population in those early years, and their contribution introduced a new lifestyle in comparison to the traditional lifestyle. Thus, a dual social structure emerged; these two components were known as the Old and New Ankara Citizens (Nalbantoğlu 1984, 258–259).

Atatürk Boulevard

Ankara was planned after the proclamation of the republic. According to the official city plan, there was one main center line along Atatürk Boulevard (Baydar 1992, 46). Ulus and Kızılay squares were two separate loci connected to Atatürk Boulevard, which has enjoyed a development process in parallel to the development of Ankara, and has been shaped by modern institutional construction. From this perspective, Atatürk Boulevard was designed to link the old city center, Ulus, to the new city center, Kızılay.

Ulus Square

Ulus Square was called Taşhan Square until the Republic era. The creation of a square in that area was due to the demand for a bazaar, the arrival of official bodies, the presence of the Taşhan building and the introduction of railways to the city. Taşhan Square constituted the image of a Westernizing empire to both newcomers and existing citizens in the city. This area, which had been the center of command and logistics throughout the War of Independence, had an outstanding identity. When the Congress was opened, the government offices became the district of ministries and the Taşhan building became the most important hotel. Taşhan Square, which witnessed the execution of opponents of the regime and various protests, gained an identity as a center of the national struggle thanks to that characteristic. Events such as celebrations and ceremonies were held on Taşhan Square (Sahil 2003, 14–17).

After the proclamation of the republic, the old city center, Ulus, gained the identity of a bureaucratic and official center in the capital city of Ankara. Such an identity emerged as institutional structures of the republic began to be shaped in parallel with this center. Buildings, such as the Second Congress, Ankara Palas and train station were meaningful for deputies and other bureaucrats who lived in Istanbul but worked in Ankara during the early years of the republic period. Ankara Palas is a type of hotel where the group called the New Ankara Citizens pursued their social activities, representing the modern life. Besides classical music sessions, other events, such as meetings, parties and balls, were held at the hotel. Classical music concerts were also given at a stage in the yard of the Second Congress Building (Bayraktar 2005, 23–24). The City Garden was organized as an outdoor area where young people danced and concerts were held. This new lifestyle demonstrated by means of new spaces was also important for women's participation in social life.

In this process, the meaning of Taşhan Square changed. After the erection of Atatürk's Monument, which became the most important visual element of the square, the site's name was changed from Taşhan Square to Ulus Square. It received a new identity as a locus of public relations and a stage for a new public life, where celebrations and ceremonies were held to express excitement with the republic. At the same time, both the Old and New Ankara Citizens met during the celebrations held in this place (Baydar 1992, 45).

Ulus started changing in the 1930s. As new banks were built in the district, led by the Central Bank, it began to gain the identity of a financial center as well. The bureaucratic characteristics of Ulus started to fade upon the construction of new ministry buildings in Yenişehir and bureaucrats moving to that district. So Ulus Square, known for its modern buildings, changed into a communication place where Old and New Ankara Citizens met during their visits to the banks and shops. However, the square's ceremonial identity was also important and it was still a meeting place for them during the celebrations of the republic.

Upon the building of the Ulus Office and Commercial Block in the 1950s, Ulus Square gained a new scale and identity. As an outdoor space affiliated with the surrounding buildings, the square was reorganized and the monument was replaced. But the new arrangement, with its emphasis on the ceremonial characteristics, did not allow citizens to rest at this place.

The 1950s marked the beginning of a new era in Turkey. The urban population increased drastically because of migration from rural areas to the cities due to mechanization in agriculture and new opportunities that cities offered. In this process, random settlements accommodating migrants and the existing formal settlements in Ankara created a dual structure in social and spatial terms. The dilemma between the Old and New Ankara Citizens disappeared as a consequence of the integration of these groups. Due to the intensive flow from rural areas to the city, a different dilemma rose due to the unequal income distribution.

Since 1950, the users of Ulus, the only commercial center where entertainment venues were located beside leading banks, offices and work places, have changed in parallel to the change Ankara has gone through. Squatter settlements surrounding Ulus were mainly preferred by large numbers of low-income groups (Altındağ Municipality 1987, 63) who gradually learnt to visit the square and to use its amenities. On the other hand, Ulus started to lose its reputation as the single main center as Kızılay, the new center, grew as a commercial focal point. So the use of Ulus Square changed and it became a contact place for medium- and lower-income citizens visiting Ulus for shopping, and a meeting place during celebrations.

In the 1980s, Ulus also lost its medium-income citizens. As those with higher incomes moved to the southern parts of the city, Kızılay turned into a new shopping center for medium-income citizens, while Ulus started to lose its importance as their preferred choice (Ayten 1997, 327–328). Those on lower incomes were shopping at Ulus, and it became a contact place for them.

This characteristic of Ulus Square is still alive today. Ulus is also visited by citizens for certain specialized commercial goods such as handicrafts, and also by those who want to visit an important religious and historical center.

Kızılay Square

Kızılay Square was called Havuzbaşı after the proclamation of the republic. Havuzbaşı, where a new lifestyle was conceptualized and represented in Ankara, was the most prestigious place in the modern capital city of the Turkish Republic (Çağlar et al. 2006, 178). There was a pool with a baroque-style group of statues which gave its name to Havuzbaşı (Bayraktar 2005, 24). The users of Havuzbaşı were the citizens living in Yenişehir district, a developing region in the southern part of the city where bureaucrats' houses were located (Sahil 2003, 18). With these characteristics, Havuzbaşı became a relaxation and communication place for the New Ankara Citizens.

When the Kızılay building was erected in the 1930s, the square was renamed Kızılay Square and in the course of time the adjacent Kızılay Park and Güvenpark replaced Havuzbaşı as the new relaxation and communication place for the New Ankara Citizens (General Directorate of Maps 1983, 48).

Upon the construction of the ministry buildings during that period, Kızılay started to gain importance as the new bureaucratic center. The New Ankara Citizens, most of whom were government officers, were taking walks on Atatürk Boulevard, which had wide and green pavements and pastry and pudding shops. This process meant the importance of Kızılay Square as a focus of recreation increased together with that of the boulevard (Bilsel et al. 1997, 3–4).

After the 1950s, Kızılay Square gained a new scale and identity upon the building of a skyscraper. New formal settlements arose around the ministry buildings and Kızılay developed. Medium- and upper-income citizens living in these settlements constituted the main group of visitors of the Kızılay area. As time went by, the residential use was replaced by a commercial one on Atatürk Boulevard and Kızılay started to be perceived as a commercial center as well. Branches of banks and shopping centers were opened in Kızılay.

On the other hand, in parallel with rapid changes in the city, streets were widened; Kızılay Park disappeared and a small courtyard remained in front of the Kızılay Building. At the end of this period, when the Kızılay Building disappeared, the courtyard was cleared away as well (Çağlar et al. 2006, 181). Thus, Kızılay Square lost its characteristic as a leisure and communication place. However, as an outdoor space linked with Kızılay Square, Güvenpark gained importance. Kızılay Square turned into a contact place for medium- and upper-income citizens.

In the 1980s, Kızılay gained the image of a center of business, commercial and administrative activities. In the southern districts, Kavaklıdere began to develop as a subsequent center (Ayten 1997, 327–328). Kızılay users who were upper-income citizens flowed out over time, and thus the number of medium-income citizens increased in the center. Kızılay Square gradually turned into a contact place for medium-income citizens coming to shop.

Though today Kızılay Square is not a square in spatial terms, it is a social place commonly used by citizens on special occasions, when the traffic flow is interrupted and it is used as a meeting area for various activities and events. Moreover, the square is a site where political meetings and protests take place. It witnesses many unofficial protests and meetings as well as signature campaigns and political events. In this sense, it is an important place for citizens from different social groups but with similar views.

Conclusion

In the capital city, Ankara, Ulus and Kızılay squares present the most important loci of experience in terms of urban life practices. Both squares have been the stages for a new contemporary life. However, meanings attributed to the squares have changed with the rapidly changing development of the areas Ulus and Kızılay, and the uses and users of the squares have evolved as a consequence of the spatial separation in Ankara:

> Until the 1930s Ulus Square (Taşhan Square) was a ceremony and celebration place for both Old and New Ankara Citizens. Kızılay Square (Havuzbaşı) was a relaxation and communication place for the New Ankara Citizens.

> In the 1930–1950 period, Ulus Square was a meeting place for Old and New Ankara Citizens who went there for banking and shopping. Kızılay Square was a leisure and communication place, together with Atatürk Boulevard, Kızılay Park and Güvenpark, which maintained their importance as sightseeing and recreation places for the New Ankara Citizens.

> In the 1950–1980 period, because of migration from rural areas to Ankara, citizens were shifted economically. Thus the uses and users of squares changed. Ulus Square became a contact place for medium- and lower-income citizens. Kızılay Square turned into a contact place for medium- and upper-income citizens.

> After 1980, Ulus Square was used mainly by groups of lower-income citizens. Those with medium incomes went to the center of Kızılay, and Kızılay Square. Upper-income citizens went to Kavaklıdere, in the southern part of the city. Ulus Square became a contact place only for lower-income citizens, and Kızılay Square a contact place only for medium-income citizens.

> Today, both squares have lost their importance for the citizens. Rather, they have started to be perceived as traffic crossroads. In the development process of the city the whole transportation concept has been addressed with an emphasis on vehicles, while pedestrians occupy a less important position. Now there is not a square designed as a meeting place where Ankara citizens can come together in this sense. However, Ulus and Kızılay still bear the characteristics of squares, both in terms of their location in the city and their uses.

To increase the importance of Ulus and Kızılay squares, new spatial arrangements are needed for them to become relaxation, meeting and communication places, so that such separation experienced in the city is eliminated. The existing potential of both squares needs evaluation. Ulus is an historical center, but today it is an area of depreciation. Reconceptualization of Ulus Square to bring its historical identity to the forefront seems to be the only way to allow citizens to be together. Kızılay is the converging point where citizens can get together. Conceptualization of pedestrian streets in the vicinity and the development of Güvenpark as an integrated body will create an important place where citizens can co-exist, and Kızılay Square will gain its former meaning and importance.

Addressing both squares and creating new spaces where all user groups can co-exist and remain in contact with each other is a necessary and mandatory contribution to enrich the social life of the city.

References

Altındağ Municipality (1987). *Construction Plan for Preserving and Developing Ankara Castle.* Ankara: Ministry of Culture and Tourism.

Araz, N. (1994). Mustafa Kemal's Ankara. In *Ankara Ankara.* Istanbul: Yapı Kredi Publishing, 347–359.

Ayten, M. (1997). Shopping Centers as a Reflection of Urban Space and City of Ankara. In *Ankara Symposium.* Ankara: TMMOB Chamber of Architects, Ankara Branch, 321–329.

General Directorate of Maps (1983). *Capital Ankara.* Ankara.

Baydar, L. (1992). Ankara-Atatürk Boulevard. In *Ankara Journal,* 4: 45–56.

Bayraktar, N. (2005). About Life and Space in Modern Ankara. In *Bulletin,* 31. Ankara: TMMOB Chamber of Architects, Ankara Branch, 23–27.

Bilsel, G., Atak, E., Gökçe, B., Sezgin, D., Şan, H. and Şişman, O. (1997). Leader-Model Role of Ankara. In *Ankara Symposium.* Ankara: TMMOB Chamber of Architects, Ankara Branch, 3–16.

Çağlar, N., Uludağ, Z. and Aksu, A. (2006). Hürriyet Square: Renovation and Transformation Story of an Urban Space. G. Ü. *Faculty of Engineering and Architect Journal,* 21(1): 177–182.

İleri, S. (1994). Two Ankaras. *Ankara Ankara.* Istanbul: Yapı Kredi Publishing, 361–363.

Nalbantoğlu, Ü. (1984). The Middle Class Group in Republic Period's Ankara. *Ankara Through History.* Ankara: ODTÜ Faculty of Architecture, 289–299.

Sahil, S. (2003). Architecture Developing and Changing in Ankara during Republic Period. In *Ankara with All Aspects.* Ankara: Ankara Metropolitan Municipality, 12–19.

Sarıoğlu, M. (2001). *Ankara: A Story of Modernization (1919–1945).* Ankara: T. C. Ministry of Culture, Pano Ofset Publishing.

Tellan, T. (1997). From Capital Ankara to an Avant-Garde City of Culture. In *Ankara Symposium.* Ankara: TMMOB Chamber of Architects, Ankara Branch, 53–61.

7 Order and heterotopia in an urban space

The case of a Spanish square

Francisco Adolfo García Jerez

The importance of urban space for the social life

We could affirm that *space* and *place* as concepts are of a peculiar and controversial notion. This is the case because of their daily and ordinary uses. For that, it seems pertinent to establish our own definition, which is based on some particular authors. Marc Augé (1993) is inclined to think more about the anthropological concept of place than space. For this author, the former refers to territory that is able to generate social relations and identifications and that possesses historicity. The place, therefore, is characterized by its physical nature but also by its symbolic component. As the space is appropriated and lived in by social groups, thus begins the process by which it is provided of new forms, uses and meanings; or, on the contrary, the existing ones are reproduced. In sum, the place, following the ideas of Michel de Certeau and Pierre Mayol (1999), is a physical space turned into a practised space. Pierre Mayol (1999) calls this phenomenon "spatial consumption". In this it converges, on the one hand, the particular privatized space, generated by the everyday use of walkers, and, on the other hand, symbolic benefits. With this he refers the configuration of tactics to the spatial appropriation. These tactics can be expressed through "discourse(s) of sense".

In the introduction of *Lugares e imaginarios en la metrópolis*, Alicia Lindón and her colleagues present four ideas of space: a naturalist one refers to the natural environment where the human being is inserted; the second is about 'absolute and relative space', in which points, lines and areas establish meaningful place, distance and regions; the third is 'produced material space', understood as those spaces created by society, depending on its technological development, its needs, institutions, and social, economic and political structure; finally, the last one refers to 'lived and thought space', and this one, according to Gumuchian, 'turns into an object of study through the meanings and values given to it' (mentioned in Lindón et al. 2006, 12).

In the present chapter we have specifically chosen to understand space as a produced material territory, lived in and thought of by social groups. It is certain that within the specialist anthropological literature we can find many kinds of these lived spaces. Each space has a distinct nature. The most famous dichotomy is between public space and private space. However, this dichotomy is not as simple as it seems. Taking into consideration property rights and management, Chiodelli and Moroni (2013) elaborated a typology of spaces depending on the degree of tolerance. Thus, the authors distinguish six types:

i   Stricto sensu public spaces: that is, public spaces for general use. These are typically spaces of a connective and open type: public squares and plazas, streets, pedestrian areas.
ii   Special public spaces: public spaces assigned to special functions. These are spaces in which more specific public activities take place, such as public schools, hospitals, libraries, playgrounds, cemeteries, and parks.
iii   Privately run public spaces: namely, publicly owned spaces that are leased to a private subject. The period of the lease may vary, but in all cases it is temporary. These include marinas, lidos, and public areas for temporary markets.
iv   Simple private spaces: namely, private spaces typically for individual/private use. This applies principally to private houses (e.g., detached houses), and more generally to all those private places where simple individual activities take place.
v   Complex private spaces: that is, private spaces in which use is conceded only to a specific group of people, usually an association or club. This is typical of various forms of "contractual communities" (such as homeowner associations, proprietary communities and residential cooperatives) or sports clubs, for example (. . .)
vi   Privately owned collective spaces: that is, private spaces that have relevance for the public, such as bars, restaurants, hotels, shopping centres, and cinemas.

(2013, 3–4)

Each one of them is characterized by a degree of right to public use or exclusiveness. Obviously, in ideal terms, these 'stricto sensu public spaces' tend to be more open, more tolerant and more inclusive than the others. Thus, with this taxonomy we emphasize the practical and symbolic use and control of public space as a tangible issue. Simultaneously, we move away from the conception – based on Habermas and Hannah Arendt's proposals – of public space as a sphere of creation of citizenship.

In any case, and starting from this dichotomy, we run into another more specific concept of space as 'contested space' (García 2011; Herzog 2004; Schmelzkopf 1995), that is to say, those spaces of a social conflict between different social actors. Another very different type of space would be the peaceful-space; it is almost always a hidden corner whose atmosphere is calm. Related to it is the idea of a 'cappuccino space'. In the opinion of sociologist Sharon Zukin (1995), these are those public places that little by little are turned into private spaces by the franchisees, or, using her own words, some public spaces suffer a 'pacification by cappuccino'. With this metaphor Zukin extends the critique, among others, of Henri Lefebvre (1972) about urban planning and its capacity to transmute spaces from 'social laboratory' into 'places of consumption'.

In any case, the majority of studies in anthropology, philosophy and geography consider that urban space is a key factor for society. Among them we highlight the anthropologist Edmund Leach (1976), who interprets the space as a cultural symbol, i.e. a product that reproduces a social particular order and displays the collective aspirations, fears and needs. In a similar vein, the philosopher Roland Barthes (1990) offers a semiotic perspective to understand the space – especially the city – as a text full of symbols and, therefore, an object with its own semantics. He called it 'the semiotic of urban discourse'. Further, the social geographer David Harvey refers to the need to take into consideration the micro – the biographic perspective – and the macro, the structural perspective (Harvey 1973). All of them tend to explain the effects of spatial surroundings on individuals.

In fact, it could be stated that the mere existence of a place that is used by various segments of society may mean a higher degree of social cohesion; at least this was one of the conclusions of Loïc Wacquant (2008) with regard to Chicago suburbs. He affirms that one of the clearest differences between the American ghetto and the poor French neighbourhoods lies in the less intensive use (or even complete absence) of public places, evident in the former in contrast to the latter. The two physical contexts belong to different categories. Therefore, community places are a significant element to consider – an axiom that was mentioned, among others, by Peter Berger and Thomas Luckmann (1966) when showing that space is a basic dimension of daily life that helps forming people's identities, or by Henri Lefebvre (1972), who defined the street as a social laboratory, a place where one learns to live together with others and to solve small conflicts. Likewise, it is a stage where society can express collectively and symbolically its discontent or joy by means of rituals. Following Henry Lefebvre, the space is not only a container which facilitates the 'biological reproduction', 'the reproduction of labour power' and 'the reproduction of the social relations of production', but it is also itself an object to be produced. This action of spatial production is formed by three aspects: the spatial practice, which 'embraces production and reproduction, and the particular locations and spatial sets characteristic of each social formation' (Lefebvre 1991, 33); the representations of space, 'which are tied to the relations of production and to the "order" which those relations impose, and hence to knowledge, to signs, to codes, and to "frontal" relations'; and representational space, 'embodying complex symbolism. . .linked to the clandestine or underground side of social life' (Lefebvre 1991, 33). Therefore, the production of space can be considered a deliberated act carried out by social sectors, frequently performed by the elite. Their actions would be fundamentally orientated by their economic, political and social interests. Starting from this premise, the anthropologist Setha Low (1996) has proposed the notion of 'space culturalization'. It converges, on one hand, the 'social production of space' (i.e. the economic, political and technical factors underlying the forms acquired by the place), and, on the other hand, the 'social construction of space'. This refers to the meanings and uses given to the places, regardless of the ones attributed by the 'power'.

In any case, urban space as public space should involve free access and be a space where individuals are considered citizens and where a considerable part of social life is produced. To sum up, this is an illustrated conception of urban space. In fact, following the ideas of William Whyte (1988) and Sharon Zukin (1995), this type of space is currently an object of an accelerated process of transformation into a space increasingly ordered and controlled by the administration: norms that regulate its use, security policies (surveillance cameras, police control) and ever-increasing private areas. This spatial process entails the gradual reduction of urban social interactions on public space.

However, urban space is still consumed by different social groups. Hence, spatial conflicts emerge regarding its control. These tense situations could be defined, according to Ágnes Heller (2002), as forms of daily friction that, although not totally free of particular feelings and interests, are principally motivated by generic, mainly moral values. Conflicts used to be located in those places in which different social groups with diverse lifestyles, different visions and spatial practices notoriously coincide. Due to this mixture they are interesting places. They are what Michel de Foucault called 'heterotopias': a heterotopia '. . .is able to juxtapose several spaces in a single real place, several sites that are incompatible' (Foucault 1986, 25). Foucault pointed out other characteristics as well. One of them is the existence of two types of hetero-topias: the *crisis* and the *deviation* one (Foucault 1999). The first alludes to places that are different from 'normal' ones and where it is possible to carry out specific social relations and roles, while the second refers to those spaces where people possess a different behaviour with regard to the conventional one. Likewise, each heterotopia has a precise role within society. Moreover, heterotopias are linked with temporal breaks called 'heterocronias'. Finally, heterotopias are capable of creating an imaginary space. Most recently, authors have reinterpreted this concept, one of the most sugges-tive proposals being made by Rodman (1992) when talking about a multilocal and multivocal place in Melanesia. Likewise, another example is the study of Chinatown in Washington, DC, by Lou (2007), where he uses this term from a semiotic point of view of town-planning processes.

However, the square analysed in this present chapter could be defined as a heteroto-pia since it assumes some of the characteristics mentioned above: it is a place where other kinds of spaces appear, even while being incompatible with them due to the fact that some users show deviational behaviour, according to the hegemonic norms. Likewise, applying urban policies such as 'touristification' (Morell 2009) and 'gentrification' (Smith 1996; Pacione 1990; Glass 1964), the local government attempted to establish certain order in the square; thus, to discipline the space through what Foucault (2006) called 'disciplinary power', that is, a set of tactics, techniques and skills whose goal is to homogenize space and its users. The homogeneity means, in this case, to foresee and control the de facto unforeseeable behaviour of people.

How and what mechanisms have been used by the local government to reach that homogeneity? In my case study the hegemonic social actors have designed and applied a policy based on the idea of making this controversial square and its surroundings attractive to tourists. Hypothetically, this policy and its consequences would guaran-tee the urban and social transformation of this geographical area. As Cameron and Coaffee (2005) suggest, culture may work as a dynamic device for the gentrification phenomenon. Thus, to live in an urban zone with a high grade of historic elements or with an intensive and attractive cultural life stimulates the settlement of new social groups, who will replace those with less economic capital. Some authors, such as Florida (2008), call these groups of artists, linked with the artistic world and prefer-ring underground zones, 'neo-bohemians'. Apart from this artistic atmosphere, one of the most relevant factors for the gentrification process has been the high and notable presence of material heritage. This heritage can become the main attraction for tourists that, in some cases, has led to a 'sclerosis effect' (Bader and Bialluch 2009; Colini et al. 2009; Morell 2009) for the geographical area where this material heritage is located. This effect frequently means, on the one hand, the gradual concentration in a specific area of touristic activities fundamentally related to leisure and 'high culture', such as restaurants, museums, architectonic motifs, and souvenir shops; and, on the other hand, the closure of local shops, the zonification of areas and the loss of inhabitants as well as spontaneous everyday life.

A square between stigma and heterotopia

The ethnographic case study chosen in this essay has as a protagonist a small square called Pumarejo, which is situated in the north-eastern part of the historic centre of

Seville, in the districts of San Luís and Alameda. For its analyses, we took into consideration the follow methodological suggestions of Setha Low:

> These approaches are appropriate for different kinds and levels of research. For instance, the individual-based methodologies (cognitive, phenomenological, and discourse) are excellent for eliciting individual users' experiences and perceptions of the site, while the societal-based approaches (historical and discourse) provide methods that uncover historical significance and social change. All of these methods answer some research problem . . .
>
> (2002, 33)

Thus, we applied three different methodologies: the observational methodologies, based on behaviour and activities of individuals and groups in a particular space; the ethnographic approaches, whose focus is an historic and social-political context analysis in order to understand the cultural spatial patterns; and the discourse approaches, in which the experience as well as reciprocal activities between individual and objects are interrelated. Thus, direct observation on selected spaces, a review of documents and newspapers, and standardized and semi-standardized interviews were the main social techniques applied from 2004 to 2009.

Pumarejo Square is in Seville, a city in the south of Spain. Seville has approximately 705,200 inhabitants and its historic centre is one of the biggest in Spain and Europe. The historic centre (with 56,733 residents) was one of the first heritage sites declared by the Spanish government in 1933 (García 2011). To return to the districts of San Luís and Alameda, the area where Pumarejo Square is located, these neighbourhoods have approximately 7,000 inhabitants and measure about 800 hectares. It is an area of small industries, established during the nineteenth and twentieth centuries, when it was one of the most important working-class neighbourhoods of Seville (García 2009). In contrast, the social classes with more economic resources live south of the historic centre, close to the monumental area with emblematic buildings representing (political and economic) power – the cathedral, the town hall and the most important squares of the city (García 2011, 2009).

The north-eastern working-class quarters of San Luís and Alameda began to suffer from degeneration at the end of the 1950s. It was physical and social negligence. It was physical because the local government did not invest in infrastructure or in restoration; it was social because middle- and upper-middle-class families started to abandon this part of the city (García 2011). Their dream was, apart from living in a space with more public services, to live in five-storey buildings with an elevator as a symbol of modernity and social advancement. The neighbourhoods of San Luís and Alameda were physically degraded and inhabited by an older and poorer population.

At the same time, this objective reality and its image attracted other groups, some with illegal or hardly respected activities. Drug-addicted people started to live in the neighbourhood. They consumed and sold drugs at points next to Plaza del Pumarejo, and prostitutes started to work in the streets next to Alameda. Additionally there were groups of young students searching for low-priced flats close to the university, and young artists and craftsmen and women who started to occupy and work in the small industrial warehouses abandoned at that time (García 2011, 2009).

This situation continued until the 1990s, when the local government undertook a survey titled 'Urban San Luís–Alameda', which gathered data on different aspects of the zone's social reality. They counted 25 brothels (with 800 to 1000 prostitutes), 300 drug-addicted people and 20 spots where they sold and consumed drugs. Regarding the population, 45 per cent did not have a complete basic education, 40 per cent were unemployed and 19.5 per cent were beyond 65 years old (seven points above the average in the city of Seville). As far as the buildings were concerned, there were 51 ruined houses, 39 vacant sites and 251 homes in a very bad condition (Ayuntamiento de Sevilla 1995). The study revealed the image of a marginal and decadent neighbourhood. This was one side of the reality, but not the only possible image of the district, since, in accordance with the exhaustive ethnographic work carried out by Cantero and others in 1999, it is necessary to stress that there was as also an important amount of young people with a high degree of dynamism and creative spirit. Therefore, the results of the survey were the local government's perfect pretext to apply for European Union (EU) funding programs in order to develop a huge plan: El Plan Urban. The plan was an initiative, funded by the EU, with the goal of renewing marginalized neighbourhoods in cities all over Europe. With regard to our case study, the Plan Urban was carried out from 1990 to 1999 with a budget of 16.24 million Euros. However, according to the analysis of Díaz (2008), one must conclude that the final results were not completely satisfactory when compared with the project's initial goals.

One aspect that is especially interesting must be stressed: the central role that the district's image played within the project. The plan started from one main axiom: in order to change the 'marginal' social dynamics of the area it was necessary to attract the middle class with more resources. This meant, of course, accepting a process of gentrification, i.e. the urban process by which the poorest residents of a neighbourhood are forced to move out and are replaced by a new and richer population. The question then was: how to attract these groups? Local government institutions used two mechanisms to realize this aim: primarily, cleaning the houses' surfaces and improving the streets' condition, which meant the physical restoration of space; and, secondly, they projected another image of the district that was going to be successfully implemented among inhabitants of the whole city. This new image, especially of the Alameda boulevard and neighbouring streets, was based on the groups of young people and their leisure activities, so that they no longer spoke about a marginal space but about a zone of *movida* (a concentration of stylish and bohemian bars); and the 'traditional' flavour of these quarters due to the existence of old streets, a traditional street market *de toda la vida* – 'for the whole life' – and small shops. It was an image of an urban postcard much more attractive than the 'marginal' and 'old' one. Somehow, the local government, together with other hegemonic forces such as construction companies, had to eliminate the area's stigma. Borrowing the concept of 'stigma' from Goffman (1986), defined as discreditable attributes attached to a person, who is being questioned as an abnormal person deprived of his/her social value, this notion could be also applied to spaces. Thus, a place, such as Pumarejo Square, may be characterized exclusively by negative attributes because of its historical, social and material decadence. This negative image does not allow the recognition of other realities, some of them composed of positive aspects. Therefore, this situation confirms that the place, such as Pumarejo Square, is also the subject of a stigmatization process. Indeed, as with any stigma, a spatial one such as Pumarejo's can be suppressed too.

As mentioned above, from the 1980s onwards and especially during the 1990s, groups of young people started to settle in the quarter, since they were looking for cheap places to live and for 'refuge spaces'. Therefore, they would colonize space in a very particular way. It should be pointed out that they did not represent all young people, but only a very particular sector characterized by those young persons who take part in collectives and associations that participate in public matters. Many of their actions can be classified as political. Some authors, such as Laraña and Gusfield (2001), have given a new name to these collectives, i.e. they form part of new social movements with close connections to the anti-globalization movement (Ayres 2004). In any case, one of their struggle's main themes is their *barrio*, the northern area of Seville's centre, a neighbourhood that has changed a lot since the 1990s. Therefore, it is a battle around the 'social production and construction of space' (Low and Lawrence-Zúñiga 2003; Low 1996). These collectives and associations use public space, but also create their own physical and symbolic places. Both of them are characterized by the existence of the same logic, since they work in a self-sufficient way (they do not receive public money) and an autonomous way (they are neither affiliated to a big organization nor to an institution), and their activities mostly have a political meaning, criticizing capitalism and the individualism of social relations.

One of these groups was the Casa de la Paz (House of Peace), formed by three associations whose primary concern is pacifism. They are associations that rent a place in Plaza del Pumarejo. You will find a library specializing in the pacifist movement in the flat, and a publishing house for alternative books (Atrapasueño, which means 'Dreamcatcher'). It is also the operational seat of the local Indymedia Estrecho: an alternative web page that connects the social collectives of the region (Andalusia) with those from the north of Morocco and with other European activists. These associations have very frequently used the 'plaza' to produce collective actions for peace.

A second group is Plataforma por la Casa del Pumarejo. It is a group of people (not all of them young) who work in order to preserve a neighbourhood centre, one of the last *casas-palacios* (palace-houses) of Seville. The *casa-palacio* is still inhabited by modest tenants who are in danger of being expelled, as the local government plans to sell it and transform it into a hotel. The neighbourhood centre has been fighting for seven years. Activists' discourse and their actions speak not only about the *casa-palacio*'s problems, but also about the gentrification process in general and about lifestyle changes. They believe that they are observing a profound change from the communal way of life ('the old neighbour') to an urban way of life ('the new neighbour') – typical for the upper-middle-class person who does not maintain social relationships in the quarter. People created a self-sufficient and autonomous centre inside the *casa-palacio*, and a lot of their actions condemn gentrification. The majority of protests are realized in the Plaza del Pumarejo (www.pumarejo.es/).

A third collective was named Casas Viejas. More than a collective, it was an assembly of young squatters, a lot more aesthetically punk than the others. They occupied an abandoned warehouse in Aniceto Sáenz Street from 2000 to 2007. Their internal logic and aesthetics are similar to squatters in any European country, although with some important local characteristics. Their discourse speaks about the district's gentrification process and about urban speculation in the city. As an alternative, they offer their own physical and cultural space with activities such as flamenco courses, theatre, a cinema, concerts, conferences and popular meals.

However, the neighbourhood with these alternative spaces, especially Pumarejo Square, was considered a dangerous and dirty place by the rest of the city. It really was a very dynamic place and not dead at all, where the different social groups brought in a variety of lifestyles that, although not free of conflict, mainly animated the public space. The daily living experiences produced what Erving Goffman (1963) called 'not focalising interactions': those forms of interpersonal communication that are the result of prolonged co-existence. To put it in Foucault's terms, it is a sort of heterotopia. This simple and – most of the time – harmonious co-existence, however, generated small violent episodes in the square that created two antagonistic visions of space and generated a spatial conflict. One vision, the administrative one, shared by a sector of the neighbourhood – especially members of some neighbours' associations and local business owners – associated the area fundamentally with negative connotations: dirtiness, danger and disorder. They did not mention the young people and their social centres or the social dynamic of the square. Therefore, their solution or proposal for the space was that of a new foundation to produce new forms of space (a contentious issue was whether to dismantle benches so that needy people could not sit or lay down); to have more social control of the space's uses, which meant more police officers; and, finally, to give the space a new meaning, to construct a square with historic and artistic value in order to form part of the city's tourist circle and thereby change its image, uses and customs. For that it was necessary to highlight and promote awareness of the monumental heritage. Hence, they thought to rebuild Macarena's wall as a place of touristic and local interest, and organize a medieval market as a promotional tourist activity, as well as developing of guide to promote a wider tourist route through the neighbourhood. This route was planned to include Arco de la Macarena, la Casa Palacio del Pumarejo, the church of San Luís de los Franceses, San Marcos, the convent of Santa Inés, a cultural market and other places of interest. The plan was, therefore, to transform a space, understood and used as marginal one, into an ordered one used by tourists.

The other point of view, defended by groups of young people and some neighbours, was that the socially and physically degraded square had to be transformed into a more sociable space by the following means: the preservation of the existent mixture of different social groupings, especially the needy people; a new spatial form mainly based on a pedestrian place; and, as one last aspect, the organization of social events by and for the neighbourhood.

Order and disorder in the square

Due to the above-mentioned conflict, a participative consultation process was initiated between the administration, neighbours and young groups to reorder the space, its uses and its meaning. The participative process called Pumarejo: Espacio de Convivencia ('Pumarejo: space of a shared living') started in 2003. The aim was to reach a compromise between the two points of view, between these two different ways of understanding the life in an urban space. The result of the consultations was that the authorities acceded to the demands to remove the benches and to install more police control, as well as the idea that the square should be included in the tourist circle of the historic centre. From a second set of suggestions, only the demand for a more pedestrian-focused space was accepted, without offering a real alternative for the homeless people. The aim of this participative process and later intervention must be interpreted as a clear attempt to transform a 'disordered', chaotic and problematic space into an ordered one with 'tourist value' and without conflicts.

Some years after the end of the construction work in 2009, it can be observed, fortunately, that urban planning is not always able to (re)order spatial consumption patterns and urban practices: the groups of young people, the old neighbours and especially the needy people are still there, with the same daily habits as before. Sociability has been stronger than planning and has survived in spite of the increased police presence and the attempt to prohibit some of the existent social activities; for example, a small, popular market that was celebrated each Saturday, and the organization of open-air concerts. Likewise, in 2009 a neighbourhood association called La Revuelta (in Spanish it means both 'curve street' and 'riot')[1] was founded, whose paradigm is based on a conception of space close to the notion of heterotopias and tacit negotiation, as well as neighbourhood cooperation as a mechanism of cohabitation and social treatments. Although the economic crisis was – and still is – a significant factor in the failure of projects based on property sectors and land rents, some spatial conflicts resurfaced, especially related to permission for private cars to access the historic centre, the removal of benches from some plazas and the construction of underground parking close to Pumarejo Square. In 2014 a day of protest against this urban policy was organized, called *Batalla contra el Zoidoparking* ('Battle against the parking of Zoido' – the name of the current mayor). However, it is relevant to highlight the opinion of the La Revuelta representative, expressed in an interview to local newspaper *Correo de Andalucía* in 2013:

> the tourism of the city is very focused on the monumental area of the cathedral . . . [The] basilica of Macarena, and the churches such as San Gil, Santa Marina, San Luís, could constitute [another] very interesting tourist route, which will reactivate the economy of the neighbourhood.
>
> (*Correo de Andalucía*, June 22, 2013;
> accessed August 19, 2014)

In this interview, together with the defence of integrating and inclusive public space (as a common good), he mentioned the possibility that Pumarejo Square and its surroundings could also form part of the tourist routes. Both arguments show an innovative vision on public space and its hypothetical renovation. This vision also shows the rejection of understanding the social life from naïve and obsolete dichotomies based on governmental/non-governmental, touristic/native, dirty/pure and order/disorder counterpoizing. In this way, the positive nature of public space is not defined exclusively due to its public or private character, its economic or social value, but because of its capacity of preserving/promoting social interactions and economic and vital dynamism as well as the inclusion of social groups, activities and uses.

Conclusions

This ethnographic example allows us to understand a point that goes beyond the social production and construction of space (Low 1996). It concerns the organizational modes of social life in urban spaces. Frequently, administrations and big sectors of society understand 'disordered' social life in an urban space as weakness, something negative and abnormal. They prefer to convert it into clean and homogeneous space through the foundation of new forms, uses and meanings, although this means to exclude social groups and deny the public nature of urban space. For that it appeals to the rational component underlying planning techniques and urban renovation in order to avoid or dismantle what is considered anomie or a deviation from the norm.

However, the case of Pumarejo is not an exception in this part of the world; some examples of this are presented by Tomic et al. (2006) in their research of Santiago de Chile. According to the authors, under the Pinochet regime the local government carried out a town-planning policy with a clear Haussmanian tendency (Picon 2003) that established a cordon sanitaire between the new spaces designed for the elite and the surrounding areas. Another example is the analysis of Brazilian cities offered by Scheper-Hughes (1993). This author stresses the role of the police in expelling homeless children from specific central places to the periphery. The goal was to restore *order* and *prestige* in these spaces, both notions understood as synonyms for *progress*. In effect, it can be stated that authorities almost never consider disorder as an opportunity to create more democratic spaces, as a basic element of social cohesion, or as the laboratory of shared living in cities that become more and more heterogeneous and fragmented – important insights that we have learned from the French thinker Henri Lefebvre.

Note

1 For more information, visit www.larevuelta.org/

References

Augé, M. (1993). *Los no lugares: espacio del anonimato. Una antropología de la sobremodernidad*. Barcelona: Gedisa.

Ayres, J. M. (2004). Framing Collective Action against Neoliberalism: The Case of the 'Anti-Globalization' Movement. *Journal of World-Systems Research, Special Issue: Global Social Movements Before and After 9–11*, 2004(1): 11–34.

Ayuntamiento de Sevilla (1995). *Urban San Luís-Alameda*. Unpublished official report. Seville: Ayuntamiento de Sevilla.

Bader, I. and Bialluch, M. (2009). Gentrification and Creative Class in Berlin Kreuzberg. In Porter, L. and Shaw, K. (eds). *Whose Urban Renaissance? An International Comparison of Urban Regeneration Strategies*. London: Routledge, 93–102.

Barthes, R. (1990). *La aventura semiológica*. Barcelona: Paidós.

Berger, M. and Luckmann, T. (1966). *The Social Construction Of Reality: A Treatise In The Sociology Of Knowledge*. New York, NY: Doubleday.

Cameron, S. and Coaffee, J. (2005). Art, Gentrification and Regeneration from Artist as Pioneer to Public Arts. *European Journal of Housing Policy*, 5(1): 39–58.

Cantero, P. A., Escalera, J., García del Villar, R. and Hernández, M. (1999). *La ciudad silenciada: vida social y Plan Urban en los barrios del Casco Antiguo de Sevilla*. Seville: Ayuntamiento de Sevilla.

Chiodelli, F. and Moroni, S. (2013). Typology of Spaces and Topology of Toleration: City, Pluralism, Ownership. *Journal of Urban Affairs*, 36(2): 167–181.

Colini, L., Pecoriello, A. L., Tripodi, L. and Zetti, I. (2009). Museumization and Transformation in Florence. In Porter, L. and Shaw, K. (eds), *Whose Urban Renaissance? An International Comparison of Urban Regeneration Strategies*. London: Routledge, 50–59.

De Certeau, G. and Mayol, P. (1999). *La invención de lo cotidiano 2. Habitar, cocinar*. México DF: Universidad Iberoamericana.

Díaz, P. I. (2008). Movimientos vecinales contra la gentrificación y transformaciones en la política local de Sevilla. Los casos de El Pumarejo y San Bernardo X Coloquio Internacional de Neocrítica. www.ub.edu/geocrit/-xcol/8.htm [accessed March 13, 2010].

Florida, R. (2008). *Who's Your City: How the Creative Economy Is Making Where You Live the Most Important Decision of Your Life*. New York, NY: Basic Books.

Foucault, M. (2006). *Seguridad, Territorio, Población*. Buenos Aires: FCE.

Foucault, M. (1999). *Estrategias de poder. Obras esenciales*. Barcelona: Paidós.

Foucault, M. (1986). Of Other Spaces. *Diacritics*, 16(1): 22–27.

García, Jerez F. A. (2009). La Alameda que te gusta. Conflictos Sociales y Planificación Urbana en torno a un Espacio Público. In *Ciudades globales, culturas locales*. Eusko Ikaskuntza, 31: 585–599.

García, Jerez F. A. (2011). Micro-conflictos espaciales y habitus político de los grupos contra-hegemónicos *Nómadas*. *Revista Crítica de Ciencias Sociales y Jurídicas*, 31: 257–276.

Glass, R. (1964). *Introduction: Aspects of Change*. London: MacGibbon & Kee.

Goffman, E. (1986). *Stigma: Notes on the Management of Spoiled Identity*. New York, NY: Simon & Schuster.

Goffman, E. (1963). *Behavior in Public Places: Notes on the Social Organization of Gatherings*. New York, NY: The Free Press.

Harvey, D. (1973). *Social Justice and the City*. Baltimore: Johns Hopkins University Press.

Heller, Á. (2002). *Sociología de la vida cotidiana*. Barcelona: Península.

Herzog, L. A. (2004). La política, el diseño y el espacio público en la ciudad de México y Barcelona. In García Canclini (ed), *Reabrir espacios públicos. Políticas culturales y ciudadana*. México DF: Universidad Autónoma Metropolitana, 267–303.

Laraña, E. and Gusfield, J. (2001). *Los nuevos movimientos sociales. De la ideología a la identidad*. Madrid: CIS.

Leach, E. (1976). *Culture and Communication: The Logic by Which Symbols Are Connected*. Cambridge: Cambridge University Press.

Lefebvre, H. (1991). *The Production of Space*. Maiden, MA, Oxford and Carlton: Blackwell.

Lefebvre, H. (1972). *La revolución urbana*. Madrid: Alianza Editoria.

Lindón, A., Aguilar, M. A. and Hiernaux, D. (eds) (2006). *Lugares e Imaginarios en la Metrópolis*. Barcelona: UAM-I/Anthropos.

Lou, J. (2007). A Geosemiotic Analysis of Shop Signs in Washington, D.C.'s Chinatown. *Space and Culture*, 10: 170–194.

Low, S. M. (2002). Anthropological-Ethnographic Methods for the Assessment of Cultural Values in Heritage Conservation. In De la Torre, M. (ed), *Assessing the Values of Cultural Heritage*. Los Angeles, CA: The Getty Conservation Institute.

Low, S. M. (1996). Spatializing Culture: The Social Production and Social Construction of Public Space in Costa Rica. *American Ethnologist*, 23(4): 861–879.

Low, S. M. and Lawrence-Zúñiga, D. (eds.) (2003). *The Anthropology of Space and Place: Locating Culture*. Malden, MA: Blackwell.

Mayol, P. (1999). Habitar. In de Certau, M., *La invención de lo cotidiano 2. Habitar, cocina*. México DF: Universidad Iberoamericana.

Morell, M. (2009). Fent Barri: Heritage Tourism Policy and Neighbourhood Scaling in Ciutat de Mallorca. *Etnográfica*, 13(2): 343–372.

Pacione, M. (1990). *Urban Problems: An Applied Urban Analysis*. London: Routledge.

Picon, A. (2003). Nineteenth-Century Urban Cartography and the Scientific Ideal: The Case of Paris Osiris. *Science and the City*, 18: 135–149.

Real Academia de la Lengua (2010). *Espacio*. www.rae.es/ [accessed August 10, 2010].

Rodman, M. C. (1992). Empowering Place: Multilocality and Multivocality. *American Anthropologist,* New Series 94(3): 640–656.

Scheper-Hughes, H. (1993). *Death Without Weeping. The Violence of Everyday Life in Brazil*. Berkeley, CA: University of California Press.

Schmelzkopf, K. (1995). Urban Community Gardens as a Contested Space. *American Geographical Review*, 85: 3.

Smith, N. (1996). *The New Urban Frontier: Gentrification and Revanchist City*. New York, NY: Routledge.

Tomic, P., Trumper, R. and Dattwyler, R. H. (2006). Manufacturing Modernity: Cleaning, Dirt, and Neoliberalism in Chile. *Antípode*, 38(3): 508–529.

Wacquant, L. (2008). *Urban Outcasts: A Comparative Sociology of Advanced Marginality*. Cambridge: Polity Press.

Whyte, W. (1988). *City: Rediscovering the Center*. New York, NY: Doubleday.

Zukin, S. (1995). *The Cultures of Cities*. Oxford: Blackwell.

8 Contested public spaces and the right to the city

The case of Cairo's historic bazaar

Wael Salah Fahmi

Introduction

This chapter tackles the contested nature of public spaces within Cairo's historic bazaar, involving official gentrification policies and local people's responses to eviction threats. On one hand, this is explored in relation to the official reimagination of the historic district as sanitised, gentrified open museum, thus seeking increased control over public space with restricted accessibility. On the other hand, the study investigates local people's reactions and expectations as they occupy and appropriate their public spaces for various community activities whilst being confronted with compulsory eviction and lack of residential security.

There is a need to make a distinction between *commercial* public spaces and *commerce* public spaces. Whereas shopping spaces are considered *commercial semi-public spaces,* locales of the consumerist culture, the current research presumes the Cairo historic bazaar to be an example of *commerce public space*, with its handicraft market and workshops, street vendors, pavement cafés and restaurants, religious and cultural events representing a mixture of overlapping multi-functional places for local people's daily encounters. For this reason, we shall describe these places as *commerce public spaces*, embodying local traditions and informal horizontal communication and networking. This will be examined in relation to the contestation created as a result of the official vision of highly regulated commercial spaces designed for global tourism consumption.

The chapter discusses alternative ideas for pedestrianisation and gentrification of the historic bazaar, combining information from interviews with primary stakeholders – local residents from the bazaar area – and with secondary stakeholders, namely non-governmental organisation (NGO) activists, policy-makers and urban planners. Issues related to local involvement in the project were raised in relation to urban heritage policy, the management of the built environment and local quality of life. In addition, recent field study investigated the situation within the area since the 2011 uprising, particularly the partial pedestrianisation of al-Muizz Street, regarded as a pilot project within the overall rehabilitation and gentrification of historic Cairo.

Conservation, gentrification and the right to the city

Historic heritage management is discussed in this text in terms of conservation planning options regarding restoration and rehabilitation as illustrated in previous studies of Cairo's historic medieval (Islamic) city (Sutton and Fahmi 2002; Fahmi and Sutton 2003). Whilst restoration of monuments focuses on certain individual, significant listed edifices, potentially neglecting other historic buildings within the surrounding urban fabric and thus resulting in a 'museum town', made for tourists rather than for residents, in contrast rehabilitation focuses on whole quarters or districts, with the cultural built environment heritage being considered part of the present everyday life of citizens.

Since the 1980s an increasing number of cities have developed pedestrianised centre districts, creating new urban consumption spaces for public activities, as noted in a study which investigates the pedestrianisation of the main street of Beyoglu in Istanbul's central business district (Dokmeci et al. 2007). Whilst the pedestrianisation of Beyoglu's main street was based on public-private cooperation, afterwards it became a market-led gentrification process with the opening of international retail outlet stores, thus contributing to the functional transformation and changed land prices within surrounding neighbourhoods.

Moreover, Gotham (2005) highlights the broader social forces and critical issues that affect gentrification, such as urban restructuring, socio-cultural changes and actions of large corporate firms in redeveloping certain heritage sites into spaces of entertainment and consumption. In this new urban landscape, gentrification amalgamates with other consumption-oriented activities such as shopping, restaurants, cultural facilities and entertainment venues, leading to an altered relationship between culture and economics in the production and consumption of urban space (Gotham 2005). Consequently, land and property prices become neither affordable nor responsive to local needs, thus restricting resources to public agencies and private investors.

This results in social mobilisations claiming and occupying the city, in terms of residents protesting against gentrification, land use changes and evictions. People have been claiming the right to the city, as noted in residential squatters and the 'renovation-deportation' movement which opposed planned gentrification. Hence, people's occupation of public spaces has turned into an urban social resistance against the official gentrification, privatisation and eviction policies. This will be examined hereafter in relation to prospects for rehabilitation, gentrification and pedestrianisation projects within Cairo's historic districts and its contested public spaces and heritage.

The historic city, context and urban problems

Historic Cairo (969–1863 AD), covering the area built up during the Islamic or medieval Fatimid, Ayyubid, Mamluk, and Ottoman periods, currently retains a prominent physical urban character and a strong social identity, with several monuments dominating its townscape, along a north-south axis. Despite the construction of two major streets cutting across the old urban fabric, al-Azhar Street in the north and Mohamed Ali Street further south, many elements of the original street pattern are still evident in the present layout and morphology of historic Cairo. The north-south al-Muizz Street, which links these two new streets, continues to be the main axis articulating the historic city (Antoniou 1998).

The historic city is generally characterised by a decaying housing stock, a lack of public spaces, and increased population density. In addition to the problem of historic monuments being misused for inappropriate storage and commercial purposes, homeless people occupied as squatters the historic buildings, which subsequently decayed due to over-occupancy and neglect. Environmentally polluting activities such as metallurgy, marble and timber workshops and storage facilities, in addition to traffic congestion, inadequate infrastructure and insufficient service provision, have contributed to the deteriorating urban fabric.

Critique of conservation and rehabilitation plans within historic Cairo

Following designation of historic Cairo as a World Heritage Site in 1979, a 1980 UNESCO plan defined six priority zones within which new development should be restricted, suggesting that conservation action should be concentrated along the main north-south spinal route which links the main monuments whilst also acting as a focus for economic activity (Antoniou et al. 1980; Abdel Fattah and Abdelhalim 1989).

According to the Greater Cairo Region master plan (1988), the development and upgrading of the historic city aimed to preserve the traditional fabric through building control regulations combined with development of public spaces in the North Gamalia and Darb al-Asfar areas (Kamel 1992). The conservation of monuments was prioritised with their re-use for various social and cultural activities. There was a need to upgrade the surrounding built environment through the transfer of wholesale, commercial and industrial activities to the suburban eastern new settlements, whilst keeping retailing and handicraft workshops within the main historic spine. The proposed evacuated areas would be used to improve road networks and create parking areas, tourist services, open spaces and community services.

The General Organisation for Physical Planning (GOPP) and Institut d'Amenagement Urbain et Regional d'Île de France (IAURIF) rehabilitation strategy (1988–1991) proposed the introduction of new public spaces in the style of European plazas, through the re-use and renovation of monuments and through pedestrianisation and traffic control measures, whilst removing various encroaching buildings. Three projects were later suggested to rehabilitate whole districts, focused on the Sayeda Zainab Quarter in the south, Gamaliya Quarter in the north and Darb al-Asfar Quarter in the east (GOPP/IAURIF 1990). This collaborative research body produced general guidelines for the improvement of the built environment, seeking to develop the northern and southern gates through the creation of a ring road around the historic city.

A 1997 United Nations Development Programme (UNDP) rehabilitation plan, which was not put into action, covered an area of about four square kilometres in historic Cairo from Bab al-Futuh in the north to the Ibn Tulun mosque in the south. The plan identified nine clusters of historic monuments and the linking streets which made up the heritage corridor, each being a primary target for rehabilitation, upgrading and conservation. In seeking to achieve a broad-based rehabilitation, the plan combined two contrasting approaches: a tourism-based and a community-based rehabilitation. The former aimed to attract investment to restore and re-use monuments for recreation purposes, services, facilities and housing, thus involving some limited gentrification. The latter would improve local residents' housing, empower their community and restore monuments' re-use, serving the community, the business sector and the tourist industry.

The 1997 UNDP plan, which suggested a very hierarchical organisation and hardly involved local people, included the pedestrianisation of the central spine along al-Muizz Street, and other one-way streets, between 9 a.m. and 9 p.m., to ease traffic congestion. A key contribution would be adaptive re-use of restored buildings and the resurrection of the old *al-fina* (outside courtyard) concept, whereby shops and workshops could extend their activities out on to the street in front of their premises.

In 2010, the UNESCO World Heritage Centre launched the Urban Regeneration Project for Historic Cairo, which aimed to foster and develop an urban conservation policy for socio-economic revitalisation and environmental upgrading, whilst creating adequate institutional capacities and technical skills and increasing awareness of heritage issues amongst authorities and local people.

Ultimately, none of the previously mentioned plans was fully implemented apart from partial pedestrianisation of al-Muizz Street, which was included in the 1997 UNDP rehabilitation plan, and the creation of the North Gamalia axis mentioned in the GOPP/IAURIF rehabilitation strategy of 1988–1991.

Nevertheless, the official proposal (2002) adopts an inconclusive strategy of slow gentrification and relocation of local activities outside the main bazaar area to suburban locations. The second phase of the pedestrianisation of al-Azhar Street is expected to be realised, following the first phase involving partial pedestrianisation of al-Muizz Street, which has led to environmental and infrastructural problems and illegal vehicular accessibility since the 2011 uprising. However, due to a lack of official action, historic Cairo's safeguarding efforts remained limited to the piecemeal restoration of a restricted number of its monuments and to a few demonstration projects (Sutton and Fahmi 2002).

Analysis of the gentrification proposals of historic Cairo (al-Azhar and al-Muizz streets)

Despite the availability of the previously mentioned proposals, the government pursued its own conservation policies between 1998 and 1999, in order to resolve the problems of al-Azhar Bridge, the heavy traffic and its serious environmental consequences on the safety of historic monuments.

After two years devoted to the restoration of the al-Azhar and al-Hussein mosques, an underground tunnel was constructed under al-Azhar Street. As a result, the area between the two mosques was pedestrianised and transformed into a new plaza which directly created access to the central spine of al-Muizz Street and Khan al-Khalili bazaar area. Traffic was diverted largely on to an inner ring road around the historic city, which initially included the creation of the North Gamalia axis outside the northern city walls as an alternative to the proposed pedestrianised al-Azhar Street. The North Gamalia axis has resulted in the partial eviction of Bab al-Nasr tomb dwellers.

The author administered a field study (2002 and 2003) which employed ethnographic techniques, including unstructured interviews with purposively sampled secondary stakeholders (planners, local authorities) involved in the Historic Cairo Restoration Programme Al-Azhar Scheme and North Gamalia Project, and with the director of a local NGO, the Association for the Urban Development of Islamic Cairo (AUDIC). More importantly, a small area survey employed direct observation and structured interviews with primary stakeholders. According to their responsiveness and willingness to participate in the interviews, primary stakeholders were randomly sampled within the area between the al-Hussein and al-Azhar mosques (50 local residents, 10 wholesale merchants, 20 retail shop owners and a number of street vendors). Three topics were identified as guidelines for open-ended interviews and for focus group discussions: general attitudes with respect to the positive and negative impacts of the project; satisfaction with environmental conditions regarding landscape, open spaces, maintenance and management levels in the area; and future expectations concerning security of tenure and threats of eviction.

Whilst the study focuses on earlier proposals for the gentrification of the area, a brief reference to the post-2011 situation (particularly within al-Muizz Street, which has nearly 21 significant monuments) was introduced as an indicator of the impact of the uprising on official rehabilitation policies. Therefore, a follow-up small area field study was conducted in 2014 with a few primary stakeholders exploring the impact of the 2011 uprising on the situation in both the bazaar area and the cemeteries. However, as a result of the lack of security since the uprising and the continuous insurgence, there was difficulty in obtaining reliable in-depth data for the analysis of the post-2011 situation, apart from a few local narratives.

Pedestrianisation of al-Muizz Street: a pilot project

Initially, the pedestrianisation of al-Muizz Street and its transformation into an open museum were assigned most of the budget allocated for the implementation of historic Cairo rehabilitation projects (nearly LE 850 million; 1 USD = 7 LE, August 2014 rates). In 2008, al-Muizz Street, regarded as a pilot project and the first stage of the al-Azhar scheme, became a pedestrian zone between 8 a.m. and 11 p.m., with goods traffic being allowed outside of these hours. This pilot project involved restoration of historic buildings, landscaping, street pavement and furniture, repairing the sewerage system and installation of LED lights. Nevertheless, one of our interviewees, a planner at GOPP, criticised the al-Muizz project, noting that '. . .it wasn't preservation but a scheme of turning medieval Cairo into a sanitized tourist district featuring inauthentic surroundings with monuments deprived of their living character.'

However, after the 2011 uprising, the state police enforcement of al-Muizz Street as a pedestrian zone has ended. Since then, the street has suffered from environmental degradation as a result of inefficient infrastructure and heavy car traffic. Apart from the partial pedestrianisation of al-Muizz Street (which was later subjected to environmental and infrastructural problems and illegal vehicular accessibility after the uprising) and the North Gamalia axis, the gentrification of the historic district was halted. This was attributed to a lack of state security within the area, political instability, social unrest and insurgence, insufficient state subsidies and a decline in tourism sector.

Pedestrianisation proposals for al-Azhar Street and the bazaar area

Planners and local authorities viewed the project as a tourist, commercial, cultural and recreational axis which would provide inhabitants of the area, visitors and traders with sustainable services and would become an attractive cultural hub for both national and international tourists. The project was concerned with landscape and architectural characteristics (ranging from the colours of facades and the tiling of plazas and promenades to types of trees, etc.), reusing existing buildings and constructing a multipurpose commercial and cultural centre, whilst providing tourist services (restaurants, coffee shops, bazaars, etc.) within vacant land.

Three axes were proposed: (1) a tourist axis as an eastern entrance to the area, with a concentration of bazaars, three-star hotels, restaurants and coffee shops; (2) a commercial axis as the western entrance to the area, characterised by a tourist market and kiosks or stalls to house street peddlers who currently occupy al-Musky Street; (3) a cultural axis between the western fence of al-Azhar University and the eastern facade of al-Azhar Mosque, which should house a second-hand books market, linked to the existing cultural centre within the restored houses of Zaynab Khatoon, al-Harrawy and al-Sitt Wassila.

Whilst the plan aims at pedestrianisation of al-Azhar Street, a network of secondary streets will be used for accessing services to the area, with parking areas for local inhabitants and shop owners within proposed multi-storey garages (at the intersections of al-Azhar and al-Mansuriyyah and al-Azhar and Port Sa'id streets). This part of the plan will allocate specific locations for loading and unloading vans serving a wholesale market in the area.

A critical view was expressed by the local NGO AUDIC, emphasising the fact that closing the al-Azhar area to car traffic would result in the decline of traditional markets and crafts which were not available in modern department stores (textiles, leather, gold and silver jewellery, perfume essences, spices and household goods). According to AUDIC,

> The proposed intervention would devastate and disperse the community. Thousands of families' livelihoods will be affected by the decision to turn al-Azhar Street into a pedestrian street. Wholesale and retail shops will lose their clients if they have to reach their destination on foot from al-Azhar Street and surrounding areas.
>
> (Hassan 2002)

Bazaar people's reactions

A majority of 80 per cent of respondents mentioned that one of the effects of the al-Azhar pedestrianisation project was the disruption of their economic structure, social ties and community networks. They were concerned about the immediate socio-economic effect of pedestrianisation of al-Azhar Street on vehicular access to storage areas within the commercial and industrial zones (the al-Musky and Gamalia areas) and within the historic axis (al-Muizz Street). They expressed the need for accessibility of vehicles through the main service road and secondary streets.

Concerning the future, official plans to relocate commercial and industrially polluting activities caused anxiety amongst workshop owners. Despite problems of noise and pollution, caused by workshops and small-scale industries, people were unhappy with alternative suburban sites, which would disrupt their socio-economic conditions and their prospective markets (established within the historic city). The survey revealed that there were a considerable number of low-income people, residing alongside the main historic axis, who were threatened with being forcibly moved out of their houses as a result of the al-Azhar land clearance plan.

Resistance to the planned eviction scheme was expressed by two-thirds of the interviewed local residents, with elderly members of the community (59 years and above) expressing more anxiety about their relocation as compared to younger age groups (between 15 and 39 years). The issue of compensation was raised in terms of who would be eligible, whether there would be enough replacement housing, and whether it would be accessible to employment and to services such as educational and social facilities.

Other respondents pinpointed security of tenure as a problem as they often had no official documents to prove their ownership of buildings, thus facing possible eviction with minimal compensation. Nearly 20 per cent of the respondents squatted into abandoned historic monuments in North Gamalia, with 30 per cent already occupying evacuated informal housing units which were less affected by a 1992 earthquake. Other long-term and low-income residents have constructed unauthorised buildings adjacent to certain monuments in order to accommodate their increasing households.

Many utilities and services (proper garbage disposal, refuse collection and a sewage system) were urgently needed in some areas, with inadequate infrastructure contributing to most respondents' dissatisfaction. Respondents mentioned the need for educational, medical, and recreational facilities and open spaces, including children's playgrounds.

Despite scepticism expressed by 40 per cent of the respondents about the government's ability to deliver appropriate services, improvements within the urban environment were invariably regarded as being the local authorities' responsibility. Primary stakeholders regarded the proposed environmental improvements, ranging from landscaped open spaces and a paved main street to a reduction of noise pollution, as serving tourism investment rather than local residents' needs.

Attitudes of wholesale merchants and retail shop owners

Wholesale merchants within the al-Musky market and the bazaar, seemed well aware of what the authorities had been proposing. They were initially interested in the project and its potential for creating urban development activities as they appreciated the government's tentative approach in seeking to test local opinion in advance of definitive action. When they realised that this was not going to happen, many of the bazaar shop owners opposed the pedestrianisation plan. However, they suggested a six-month trial to ascertain the possible damage to the local economy of the closure of al-Azhar Street to vehicle traffic and the introduction of electric cars for shoppers. Interviewees proposed closing al-Azhar Street just for two or three days a week, so as to preserve the area's commercial role, as they expressed worries that the al-Azhar mosque and its environs would become just a tourist area, losing much of its local Cairene clientele.

The situation was more stressful for those who have set up temporary stalls, as they feared losing their source of livelihood, this being attributed to the local authority's reluctance to regularise their enterprises, a process which involves a complex procedure full of bureaucratic delays and considerable expense.

New activities expected to be introduced to the al-Azhar area

Much of al-Azhar Street could remain a road for vehicles, but strictly for access in and out of this central part of historic Cairo (which has been taken over by a tunnel and by a proposed third metro line). Only the central section of al-Azhar Street ought to be pedestrianised, incorporating the junction with al-Muizz Street, open spaces between the al-Azhar and al-Hussein mosques and narrow streets around the Khan al-Khalili bazaar area.

Furthermore, as authorities delayed plans for closing al-Azhar Street, partially or completely, as a result of public opposition and post-2011 insurgencies, it could be anticipated that a degree of gentrification might occur along this east-west axis, as suggested in Figure 8.1 and Table 8.1 (HCSDC 2002).

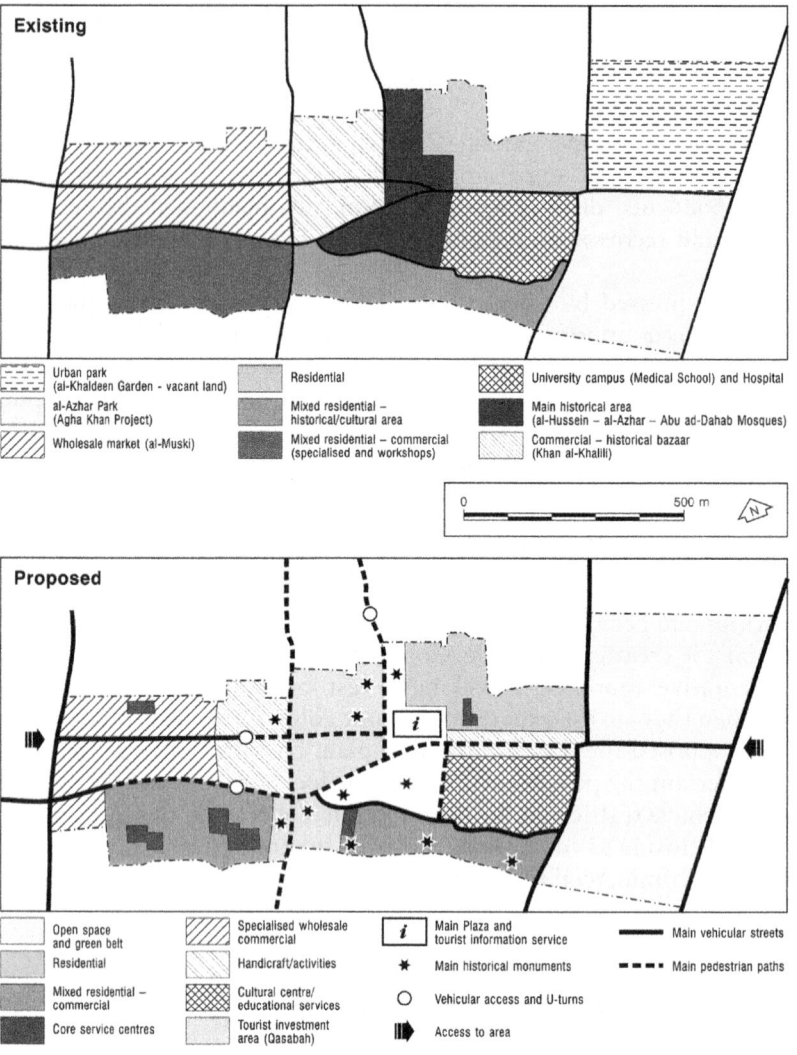

Existing

Urban park
(al-Khaldeen Garden - vacant land)

al-Azhar Park
(Agha Khan Project)

Wholesale market (al-Muski)

Residential

Mixed residential –
historical/cultural area

Mixed residential – commercial
(specialised and workshops)

University campus (Medical School) and Hospital

Main historical area
(al-Hussein – al-Azhar – Abu ad-Dahab Mosques)

Commercial – historical bazaar
(Khan al-Khalili)

0   500 m   N

Proposed

i

Open space
and green belt

Residential

Mixed residential –
commercial

Core service centres

Specialised wholesale
commercial

Handicraft/activities

Cultural centre/
educational services

Tourist investment
area (Qasabah)

i Main Plaza and
tourist information service

✱ Main historical monuments

○ Vehicular access and U-turns

➮ Access to area

Main vehicular streets

Main pedestrian paths

Figure 8.1 Actual and potential land use along the al-Azhar axis (Fahmi and Sutton 2003)

Table 8.1 New activities expected to be introduced to the al-Azhar area according to official plans

Northern side of *al-Azhar* Street

- Business centre and new office-administrative development
- Newly refurbished housing stock (and perhaps some newly-built structures) for upwardly mobile urban professionals
- Tourists' open museum with a commercial centre for marketing handicrafts
- Investment centre and specialized commercial center
- Potential commercial district with both traditional wholesale and newly introduced activities.

Southern side of *al-Azhar* Street

- Business centre and new office-administrative development
- Cultural hub
- Main historical core and tourists' activities – plaza and open museum
- Potential mixed use and new investment development area
- Main commercial district with mixed residential functions (gentrified at the expense of traditional wholesale markets)
- Business-orientated service area

Fahmi and Sutton (2003).

The historic bazaar in the post-2011 era

Since the 2011 uprising, many tourists have refrained from visiting Egypt, leading to the closure of many shops in the bazaar area; tourism industry revenue for 2011 recorded a 25 per cent fall against the previous year. Recent interviews revealed the crisis confronting the bazaar people, as noted in their various narratives:

> After the 2011 uprising, we experienced more economic hardship, with the bazaar suffering environmental degradation as a result of lack of maintenance and inefficient infrastructure.
>
> (40 years old, male wholesale merchant)

> We thought that the 2011 uprising will improve our living conditions, but after three years the situation is getting worse. I hope the government would support us by consolidating the security situation despite the fact that they [officials] want us out of the area.
>
> (61 years old, female householder)

> We have demonstrated a few times in al-Azhar Square especially after the Friday prayers so the government would recognise our demands. All we need is more security back in the area so we can work safely and so tourists would return.
>
> (28 years old, male street vendor)

> I can hardly recall a single tourist coming into my shop recently. Nevertheless, I was already struggling before the uprising as the number of tourists declined due to the global economic downturn. Therefore I asked my employees to work fewer days so I could afford to pay them.
>
> (54 years old, male retail shop owner)

However, similar causes that have motivated the post-2011-uprising protests – ranging from political repression, a lack of democratic freedoms and unsatisfactory neo-liberal economic policy reform to poverty and high unemployment rates – must all be taken into consideration when assessing the bazaar's future. Nevertheless, dispossessed social groups, such as the homeless, tomb dwellers and street vendors, may be more vulnerable to dealing with the post-2011-uprising uncertainties and difficulties that have characterised their existence for decades.

Reimagining the future of Cairo's historic district

Since the 2001 uprising, no major district development has occurred within historic Cairo, except for individual facade decorations and street landscaping. Such initiatives are mainly administered by the National Organization for Urban Harmony, which is affiliated to the Ministry of Culture. This might be attributed to various developments influencing the overall future of the area.

First, since historic Cairo is located in proximity to the main commercial spine within downtown Cairo, close to Tahrir Square, frequent insurgence – ranging from riots and rallies to sit-ins – discouraged the government from implementing various stages of the project. Second, in defiance of the main objective to create a touristic centre to serve those business entrepreneurs, the project was halted after the 2011 uprising. This was attributed to the decline in the international tourism industry, and a lack of security, which had more adverse effects on the maintenance of various historic buildings within the area. Third, for the last three years, local coffee shops have turned into venues for political public dissent and street youth activism, thus contributing to the tightening of security measures and setting up of checkpoints after midnight for both late-night strollers and street vendors. Despite the recent introduction of security measures to curb urban youths' street activities, there were various confrontations between security forces and urban youths gathered in street coffee shops. This probably discouraged the municipality from developing more public spaces with cafés, and the street markets.

Consequently, the study proposes future scenarios regarding the city as multi-functional layers in the context of post-2011 events (Table 8.2) as a result of:

- Reimagining historic public spaces in the bazaar area and cemeteries as part of official heritage, tourism and economic interventions with strict security surveillance; and
- Local people occupying public spaces and reclaiming their right to the city, as they squat in vacant monuments (as opposed to the official adaptive re-use of historic buildings) and as they appropriate streets for informal commerce activities (as noted in the case of street vendors and local markets).

Table 8.2 highlights the relationship between various developmental approaches, stakeholders and heritage management policies in relation to future scenarios. This might lead to the production of multi-layered public spaces within the historic district, involving both official views and stakeholders' right to the city in the form of:

- The residential public spaces (North Gamalia), which will integrate people's socio-economic and cultural characteristics and the built environment. This will be created through the adoption of a rehabilitation process;
- The commerce public spaces (Khan al-Khalili Bazaar and the al-Musky wholesale market), which will incorporate peoples' appropriation of public spaces such as local markets and street vendors, whilst restoring historic monuments and introducing infrastructure development, this being attributed to a process of renovation and renewal approaches;
- The heritage public spaces (al-Muizz Street) will take the form of an open museum with monuments being restored and the built environment being upgraded. As a result of a gentrification process, local residential uses will be replaced by tourism activities and facilities.

Table 8.2 Future scenarios of multi-layered public spaces (Fahmi and Sutton, 2003)

Developmental Approaches

Socio-Economic and Cultural Development			*Physical Redevelopment*		
Cultural Identity	Social Lifestyle	Economic Activities	Infrastructure Development	Upgrading Residential Environment	Restoration of Monuments

Stakeholders

Local Communities	Government/ Public Sector	Private Developers	Non-Governmental Organizations	International Agencies

Future Scenarios

Short Term Scenarios		*Long Term Scenarios*	
Renovation	Restoration	Gentrification	Rehabilitation

Reimagining the Historic City as Multi-Layered Public Spaces

Residential public spaces	Commerce public spaces	Heritage public spaces

Future prospects of an ongoing gentrification strategy

The government's approach has been regarded as too 'tourism-oriented' and so has ignored local attitudes, which could lead to 'Disneyfication' of the historic city, turning it into an open museum whilst opening the possibilities for increased land prices and property speculation. Gentrification appeared to prevail over upgrading, aided by the official plan to move people out of the historic city, causing land use changes from residential to touristic purposes.

As decayed buildings are expected to be demolished, except for registered buildings, vacant plots will permit increased property speculation within the historic district. Accordingly, future scenarios predict a slow process of gentrification and land use change within Cairo's bazaar which will enforce residents and merchants to move out to suburban districts. This involves an upward filtering of housing to the benefit of the more upwardly mobile middle classes through total renovation. The alternative model would be the rehabilitation and re-use of older vernacular buildings to create conditions for the present local inhabitants to remain, including some new younger residents, students and urban professionals living in rented dwellings.

As decrees prohibiting building demolitions were not applied, land and property speculation prevailed within the area, especially for buildings not recognised by the Supreme Council of Antiquities as heritage. The only obstacle to such speculative operations is the difficulty of evicting long-term tenants residing in some rundown apartment blocks, despite a new rental law. Consequently, some owners of residential buildings intentionally vandalise their own properties in order to have them declared architecturally unsound.

Rehabilitation should consider the socio-cultural aspect of the historic area and should not primarily be economic-development-oriented, with tourism and a Disneyfication kind of renovation programme only for the monuments. In addition, whilst the prohibition against adaptive re-use of some buildings was considered a major barrier to community participation, there remains a need to introduce local activities, commercial uses and tourist functions to help perpetuate the built environment heritage and to avoid a second phase of decay and dereliction. Previous studies of people as active participants in the making of place identity for conservation areas in Singapore (Yuen 2005), and the cooperative movement of public and private sectors in revitalising and pedestrianising Beyoglu's main street in Istanbul (Dokmeci et al. 2007), suggest that community involvement in the rehabilitation process can be maximised through establishment of conservation-oriented committees.

The study revealed that no strategic master plan has been developed and endorsed by Egypt's highest legislative and executive authorities, beyond the general objectives of conservation and enhancement of the architectural and cultural heritage contained in Cairo's historic city. The lack of an integrative framework to give structure and coherence to the range of public and private activities that could be initiated in the historic districts is clearly reflected in the array of unrelated projects and ad hoc initiatives sponsored by international and bilateral organisations and donors (Serageldin 1998).

This brings forth even more strongly the need once emphasised by UNESCO (2003) for co-ordination amongst the various institutions involved in the rehabilitation of historic districts, whilst stressing the significance of adopting a comprehensive vision, embodied in an institutional master plan which would ensure better management of the area, protecting the national heritage and taking into account the needs of the local residents.

Acknowledgement

This chapter is based on collaborative research with Keith Sutton at the School of Environment and Development at the University of Manchester, United Kingdom.

References

Abdel Fattah, K. and Abdelhalim, A. I. (1989). The Rehabilitation and Upgrading of Historic Cairo. In *The Aga Khan Award for Architecture, The Expanding Metropolis: Coping with the Urban Growth of Cairo*, Vol. 1. Conference papers.

Antoniou, J. (1998). *Historic Cairo: A Walkthrough the Islamic City*. Cairo: American University in Cairo Press.

Antoniou, J., Welbank, M., Lewcock, R. and El-Hakim, S. (1980). *The Conservation of the Old City of Cairo*. London: UNESCO.

Dokmeci, V., Altunbas, U. and Yazgi, B. (2007). Revitalisation of the Main Street of a Distinguished Old Neighbourhood in Istanbul. *European Planning Studies*, 15: 153–166.

Fahmi, W. and Sutton, K. (2003). Reviving Historic Cairo through Pedestrianisation: the al-Azhar Street Axis. *International Development Planning Review*, 25: 407–431.

GOPP/IAURIF (1990). *Enhancing Public Spaces of Islamic Cairo, Final Report*. Greater Cairo Region Long-Range Urban Development Master Scheme. Cairo: Ministry of Development, New Communities, Housing and Public Utilities.

Gotham, K. (2005). Tourism Gentrification: The Case of New Orleans, Vieux Carré (French Quarter). *Urban Studies*, 42: 1099–1121.

Hassan, N. M. (2002). Pedestrianising Sharia Al-Azhar? A Modified Project Protecting Community Livelihood. Paper presented at the International Symposium on the Restoration and Conservation of Islamic Cairo, February 16–20, Cairo.

Historic Cairo Studies and Development Centre (2002). Unpublished report presented at International Symposium on the Restoration and Conservation of Islamic Cairo, February 16–20, Cairo.

Kamel, S. H. (1992). A Master Plan Formulation in Cairo Metropolitan Area. Paper presented at the International Conference on Urban Development Policies and Projects, October 19–23, Nagoya, Japan.

Meyer, G. (1988). Manufacturing in Old Quarter of Central Cairo. *URBAMA, Elements sur les Centre-villes dans le Monde Arabe Fascicule de Recherches*, 19: 75–90.

Serageldin, M. (1998). *Notes on Strategy Guidelines for the Rehabilitation of Historic Cairo*. Paris: World Heritage Centre.

Sutton, K. and Fahmi, W. (2002). The Rehabilitation of Old Cairo. *Habitat International*, 26: 73–93.

UNDP (1997). *Rehabilitation of Historic Cairo. Final Report*. Cairo: UNDP Technical Co-operation Office.

UNESCO (2003). *World Heritage Papers: Identification and Documentation of Modern Heritage*. Paris: UNESCO.

Yuen, B. (2005). Searching for Place Identity in Singapore. *Habitat International*, 29: 197–2.

Part 3

Management and governance

Transformation and control

9 The meaning of public space in the context of space-time behaviour in the 'network city'

From socialist to sociable public space

Anastasia Moiseeva, Remon Rooij and Harry Timmermans

Introduction

Nowadays the definition of 'public space' in an urban context has shifted from its traditional meaning of streets, squares, and parks to the new collective space with programmatic functionalization. Designers, urban planners, and policy-makers are trying to re-think and re-design public space according to these new forms of sociability. This raises the concern of whether traditional public spaces are still operating spaces of the contemporary city. Has the public become so diverse that a space that everyone can identify with no longer exists? Can other places be found to replace this lost space?

This chapter therefore aims at reviewing recent transformations and reorganization of public space in post-socialist cities in the Russian Federation shifting from top-down planned policies conducted by government to bottom-up management by private investors. In contemporary urban theory there are several books analyzing and explaining in a complex way the general features of socialist cities and their transition into post-socialist ones (Andrusz, Harloe and Szelenyi 1998; Enyedi 1998; and others). The majority of existing studies deal with the economic transition, the social transition, the transition in governance or changes in the urban structure in individual post-socialist countries or cities.

We are interested in evaluating and (re)defining the spatial and social interrelations and the meaning of public spaces in the context of the new social and spatial patterns of the network society and space-time behaviour in the network city. In this chapter we make some steps in this direction by discussing contemporary trends and reflecting on some of their implications for emerging new public space and spatial urban structures on the basis of an analysis of the cultural geography of the network city.

First, the chapter reviews the inherited features of post-socialist cities from socialist times and their recent transformation towards the 'space of consumption', considering the basic approaches to study cities in urban sociology in relation to the changes in the organization of urban space. Second, the phenomenon of the spatial model 'city as network' is described. Primary focus is given to the specific characteristics of space-time behaviours: how the 'network society' appropriates and perceives public space. Third, we give a 'partial' answer to the question how to establish preconditions for the design of public space on the basis of an analysis of the cultural geography of the network city. Thoughts are given how to develop a new perspective of cultural exchange and integration of different mobilities; in which way different places relate to each other.

Finally, the chapter concludes with design strategies for the post-socialist cities where mobility is seen as one of the main sources for design preconditions.

Urbanization patterns

Urbanization patterns of socialist cities: spaces of collectivism and spatial fragmentation

The post-socialist cities have distinctive features inherited from the Soviet time. Cities were understood in terms of zones, layers, centres and peripheries. According to Aleksandrowicz the socialist city essentially consisted of (1) the main centre with vast and open public spaces: a central square with the statue of Lenin and big boulevards designed to convey the political and ideological distinction and to be a place for manifestations; (2) the place of production and work; and (3) the dwelling-place (Aleksandrowicz 1999). The standardization and rationalization of cities' basic functions (housing, work, leisure and circulation) would impart a 'mechanical unity' to the city which would become clearly bounded. The planning principles included separation of industrial and residential functions (mono-functional zones established by land use planning), concentration of commercial activities on circulation nodes and manipulation of zonal densities to achieve rational circulation and mobility patterns.

The legacy of the socialist experiments in planning and housing provision has created excessively high densities in the urban peripheries compared to the traditional market city. These areas contain a high proportion of system-built housing (mass construction) and some basic social services (schools, medical care, etc.), but a lack of sufficient retail and employment opportunities. This segregation of land uses contributes to excessively long commuting trips to the centre. Mass-produced socialist housing in the peripheral housing estates is the home of 50 to 60 percent of the population in large urban centres. Many of these peripheral housing estates are characterized by a lack of qualitative urban public facilities, and a lack of urban public life.

The socialist regime represents an example of conscious proactive attempts on the part of the state to construct new spaces for new societies epitomizing a top-down approach to build a city and society. For example, the centres of traditional 'market' cities which have developed as a result of continuous economic processes are obviously marked by mercantile factors. 'Centre' in this context means above all 'shopping' and 'business centre'. The aim of the socialist city planners was (in a contrast to the traditional 'market' city) to implement in the city centre a new ideological character. That is, instead of the market-place, the centre had to be re-organized by political institutions and by official ideology (Kostinskiy 2001).

Aleksandrowicz argued that what socialist cities had in common was an egalitarian, anti-individualistic monotony of the urban space; human individuals were integrated into a spatial order of public space which widely determined their behaviours, leaving individuals with very few alternatives for individual choice, self-expression and creativity (Aleksandrowicz 1999). Public space was a zone of collective performance. Although many Russian cities had a lot of public spaces (in terms of surface), it seems – at least at first glance – that only few of them functioned as public domains, or just very temporarily.

Urbanization patterns of post-socialist cities: spaces of consumption and spatial decentralization

Since the fall of the 'Iron Curtain', a new type of public life is continuously emerging under the rules of market economy. This new public life brings profound changes in the forms and meaning of public space. The city centre with monumental architecture, big square and the statue of Lenin, which originally carried a political and ideological message, lost their conceptual meaning – the place of manifestation became 'lost space'; at the same time new space, 'space of consumption', started to flourish.

The 'post' in the term 'post-socialist city' refers less to the past than to a new spatial 'cohabitation': post-industrial agglomerations of space and new elite residential areas; Escada and Dior stores and statues of Lenin. The post-socialist Russian society is facing simultaneously at least four types of transformation, causing a complex of structural changes: (1) political: from a totalitarian to a democratic society; (2) economical: from the planned to a market-based economy and from supply to a demand-driven economy; (3) developmental: from an industrial to a post-industrial (service) economy and society; (4) global: transformation from an isolated to an integrated position in the world economy, which is itself transformed from an international to a global type (Petrovic 2005). With the end of the 'shortage economy' and the beginning of a consumer revolution (Davis 2000), the exploitation of niche markets becomes critical. 'Space of consumption' is an inevitable outcome of pursuing commodity-oriented city development after the 'opening' of society. The townscape in recent years of reform is gradually facing the emergence of consumption space: chains of stores, supermarkets, shopping centres, malls and business parks.

Consumption is glorified and glamorized in Russia through promotional strategies and globally branded images. These new spaces are carefully designed to stimulate a sense of a new way of life. The new spaces are constructed to target those who can afford them. However, they are often in sharp contrast to their surroundings and city peripheries. A few blocks from the city centre you will find an ocean of monotonous mass-constructed houses and 'traditional' socialist public spaces, public in terms of free-accessed open-air surfaces and 'un-public' in terms of their uses. If we take a look at the contemporary post-socialist cities we can see they are simply assemblages of disconnected parts: luxurious shopping avenues and empty public squares; big-scale entertainment facilities at the peripheries and neglected public spaces in peripheral housing estates.

Urbanization patterns: globalization trends

Indeed, the transition from state-organized top-down management to commodified consumption has brought significant changes within public space and spatial distribution of facilities. The forms described above are not new and unknown for the rest of the world. The emergence of consumption spaces is influenced by the globalization trends, changing the post-socialist city from a place of work and life-as-production to a source of leisure and amenity through increasing aesthetization and functionalization of urban space. Globalization is understood as a socio-cultural phenomenon driven by the space-time compression effects and the 'dissembling' effects of the application of new transport (motorways, air transport, containerization, high-speed rail systems) and information and communications technology, or ICT (computerization, telephones, wireless technologies, satellites, optical fibre cables, etc.) (Burgess 2007).

In social studies there are three theoretical approaches that can be recognized according to their basic presumptions about what is ultimately driving socio-economic, political and cultural change and consequently changes in the organization of urban space (Burgess 2007). The first approach is based on 'technological determinism', related nowadays with the development of ICT; the second approach prioritizes the socio-economic and political explanations, while the third offers an explanation for the phenomenon in cultural terms. In reality all of these aspects are interlinked in the city, but in the frame of this chapter technological innovations (ICT development) and cultural phenomena will be more specifically addressed in relation to the recent transformations and future organization of public urban space in post-socialist cities.

In post-socialist cities the urban structures and forms inherited from the earlier Soviet period, which are fragmented and splintering, and increasing spatial segregation and dispersed activities, plus the outcomes of recent years, generate contradictions between the city centralities and, hence, become dysfunctional for the requirements of the 'network society'. At this point we should look critically at the future of public space and recognize trends which are shaping and will shape social relations and cultural practices. Referring to Castells (1996), the principal technical elements determining the nature of social, economic and cultural organization are the new transport, information and communications technologies which are increasingly integrated into global networks. This process has a dramatic effect on the organization of space and gives rise to the dominant spatial model of 'city as network'.

The network city model

In the wake of the twenty-first century, the so-called network city model is an advanced survey of the contemporary city. There is a tendency among social theorists to interpret the concept of 'network city' as cities with overlapping sets physically connected by transportation systems and virtually connected by telecommunication systems of activity spaces (Castells 1996).

Today the principal technological elements determining the nature of social, economic, cultural and spatial organization are identified as the new transport and information and communications technologies increasingly integrated into global networks. It is not just a matter of society getting 'tooled up' with the latest ICT and transport technologies but rather about societies becoming reconstructed as networks, which also effects the organization of spaces in the cities as well.

As Boelens (2000) points out, in today's 'network society' physical, social and virtual networks have taken a prominent position. These networks do not only determine our daily life in an increasing degree, but they also influence economic, cultural, political and geographical relations. Therefore, the present layout and organization of the city is under increasing pressure. The design ideas and the instruments of the majority of today's urbanists still result from the traditional geographical concept of space and time. Therefore, they are hardly suitable to play along with the fundamental fleetingness and infiniteness of the network society (Rooij 2005, 101).

Space-time convergence and new concepts of space and time

It is important to focus on the specific spatial adjustments to the technologies of the network society. The ability of these technologies to 'shrink distance' and 'stretch time' permits a dramatic increase in the volume and intensity of flows of information, commodities, finance and people on a global scale. Consequently, it changes the spatial scale at which human activities become possible and the relative significance of particular scales emerge in everyday life.

The behavioural studies point at two correlative changes: (1) individuals not only minimize travel time, but they may also value that time positively when they can visit one or more activity places which provides a higher total utility; (2) the cost of travel time appears to reach, maybe after some minimum acceptable threshold value, a maximum acceptable value. An individual's time preferences affect the morphology of that individual's action space (*action space* refers to the context, the spatial environment, in which the performing of activities and the travelling takes place – see Rooij 2005, 83). The approach of studying human spatial behaviours in a spatio-temporal context provides insights into the functioning of urban areas and public spaces, and into their spatial structure.

Space-time behaviours in the network city: an 'archipelago of enclaves'

Obviously in recent decades urban society has changed radically in a spatial sense: with the increase of motorized mobility the action space of people has increased enormously. Thus not only their spatial reach has increased, but also the diversity of activities and travel patterns. The new urbanity reflects the new, highly dynamic 'time-space patterns' of citizens: increasing flexibility in the world of employment, changes in the form of personal relationships, shared responsibilities in the home, cultural trends and recreation.

In a network society, the increasing fragmentation of individual activity and travel patterns has a major impact on visitors' use of places and spaces throughout the day (Bertolini and Dijst 2003). Every person has his or her own action space, both physically and electronically; his or her own territory where he or she comes and goes and participates in activities. These specific combinations of activities are affected by individual lifestyle, personal needs and constraints: his or her position in social networks, his or her stage in the life cycle, his or her spatial location relative to potential trip destinations, and so on. As a consequence, it appears that different, overlapping 'virtual cities' are developing, giving rise to a multiplicity of urban forms and centres. These virtual cities include multiple spatial and temporal scales of different individual and social groups. According to Bertolini and Dijst (2003), virtual cities do not have physical and administrative borders, but rather are a specific combination of activity places connected by transport networks, within definite socio-economic and behavioural constraints. This means that different groups in society follow different paths through space and time: contemporary 'spatial segregation' exists in the fact that today the spatial networks of certain groups barely overlap.

Space-time behaviours in the network city: place as a consumer good

In recent years Russians have shown an unprecedented increase in their interest in deliberate consumption of places and events. That is a consequence of the expansion of middle-class layers, particularly in the central big Russian cities. Nowadays well-to-do middle-class groups are characterized by both mass consumption and increased mobility. This phenomenon has mushroomed recently, and concerns the desire of 'ordinary' citizens to have 'interesting' lives. In the field of cultural geography these trends can be identified as 'conscious consumption of cultural experience' and 'place as a consumer good' (Hajer and Reijndorp 2001).

We can see these days that all spaces of consumption are functionally programmed in order to stimulate particular behaviours. The high priority in design is given to dictating a certain layout to public space for certain social groups who can afford it. There is a discussion in the social sphere that this kind of 'consumption parochialization' of public space, its re-appropriation by or for certain groups, can be seen as one of the most important causes of the decline of the public space as meeting place (Hajer and Reijndorp 2001, 85). Indeed, certain social groups which are considerably less mobile, like senior citizens, children from less well-to-do backgrounds, and low-income households cannot afford it and, hence, are socially excluded. It means that virtual cities of certain groups barely overlap these days. Evidently, this middle-class culture and increasing fragmentation of individual activities leads us to look at space differently: not only to pay attention to the new spaces that are created for mass consumption, but also to try to perceive and understand the way in which individuals assemble their own 'virtual cities' from a variety of elements and locations.

Key principles for public space in the network city

How to deal with public space in the network city?

The urbanized landscape of network city has now become and is most easily understood as an 'archipelago of enclaves', where everyone creates and puts their virtual city together. However, the question should not be how to bring back this transformation into an archipelago but rather what possibilities this new spatial and social reality offers for the creation of new, interesting forms of public space. Individual perception and the individualism of using public space offer a different perspective for the creation and the design of public space in the network city.

As we have already mentioned, in the network society everyone puts together their own city. Naturally these touches of individual cities, so-called 'connected interfaces', can be seen as an essence of the new design concept of public spaces. In our opinion potential urban experience will occur at the boundary of connected interfaces. The key to generate urban experience is by bringing a number of elements that are meaningful for different groups into close proximity with one another, which implies that different means, other than architecture, can be employed to create them.

The key principles of how to deal with public space in the network city are defined below. In order to gain a better understanding of the cultural possibilities of the diversity of places, we must give more thought to the way in which spatial coherence of the network city is experienced. We define two levels of design connected to network thinking: (1) *design at the city level* – to achieve spatial coherence and complementarity at city level; (2) *design at the local level* – 'space as a system of places'. These two levels of design are directed to create complementary nodes and sufficient links between them.

Design at the city level: the network thinking approach and multinodality

The traditional city is under pressure, and ideas for sustainable spatial development have to be distilled from network thinking instead of zonal thinking, from integral instead of sectorial design and planning approaches (Rooij 2005, 17).

In the new spatial model that emerges (the network city model), the principal structural elements are the spatial nodes interconnected by selected infrastructure that guarantee the all-important flows, exchanges and mobility. The most important rule is that these spatial nodes (for example, land use zones or public spaces) are configured according to their complementarities and the synergies they generate.

The concept of the network city more than ever acknowledges the possibility of social dynamics occurring consequently in the peripheries and between them. The city is no longer the monocentric conglomeration of functions, but it has grown into a multinodal urban field, the multinodal or polycentric city. *Multinodality* refers to the presence at the specified level of scale of more than one concentration of collective activities that pertain to that scale (Jacobs 2000). The polycentric structure of the city opens up possibilities of attracting a higher level of urban public and commercial facilities in marginal districts, to transform them from neglected areas to urban public spaces. This expanded network city implies a new agenda for the design of the public space, not only in the urban centres or in the new residential districts, but especially in the ambiguous areas in between.

Design at the local level: 'space as a system of places' and integration of different mobilities

We owe to the humanistic geographers such as Yi-Fu Tuan and Edward Relph the deep conceptual distinction in thinking of 'space' and 'place' (Tuan 1977; Relph 1976). Places are, for example, associated with real events (which have occurred there), with myths, history and memories. In this context, the way in which different places are related to one another in a physical, spatial and social way is a significant quality. In order to stimulate urban experiences in public space we must look for the cultural meanings of these spaces.

The answer does not pertain so much to the actual layout of separate spaces, but rather it is about the necessity of conscious design of different spaces and their interrelationships: the creation of interfaces between different 'landscapes'. This does not mean that the design and layout of those separate spaces is unimportant. However, today the design of new public spaces seldom focuses on the interfaces or intersections. In more general terms, it is important to think about the general urban planning conditions under which cultural and social exchanges occur and flourish, beyond the mere implementation of the urban programme. Therefore, the design of the 'space as a system of places' should integrate different mobilities which do not exist only because of the need to travel, but also because of the presence of possibilities to travel.

The above leads to a new social focus of the urban design laid in the reintegration, or at least realignment, of different mobilities: more than ever before it should be focused on the transitions, the crossings, the connections and the in-between spaces. This important issue demands the attention of designers and urban planners with regard to the inclusion of different worlds: civic and commercial places; public and private transport; pedestrian zones, car zones and bicycle lanes; external and internal spaces. This fluid life of urban mobilities also brings a challenge to re-think and re-design the place of the car in the city in relation to other types of mobilities.

The new meaning of accessibility and 'experienced time'

In order to understand the meaning of public space for the network society it is also necessary to introduce conceptual terms that can articulate correctly certain qualities of spaces in the network city. Here we would like more explicitly to refer to the meaning of *accessibility* and *experienced time*.

In broader terms, accessibility is not just a feature of a transportation node (how many destinations, within which time and with which means of transport can be reached from an area?), but also of a place of activities (how many and how diverse are the activities that can be performed in an area?). In this wider connotation an accessible mobility environment is thus one where many different people can come, but also one where many different people can do many different things: it is an accessible node, but also an accessible place (Bertolini 1999).

The cultural-geographic approach exchanges the terms 'space' and 'distance' with the terms 'movement' and 'experienced time' (Hajer and Reijndorp 2001). Experienced time can be understood as a new key quality in the design of public space: when everyone is creating an individual virtual city it is precisely in the experienced time. Hajer and Reijndorp argue that in future the quality of public space will be measured not merely in terms of space accessibility, but will increasingly become a question of how it influences the ambience of specific places (Hajer and Reijndorp 2001).

Otherwise, in order to create 'space as a system of places' by means of creating preconditions for overlapping interfaces between different 'landscapes' and integrating different mobilities, first we should give more thoughts to node and place quality for a given location in relation to the specific *temporal, spatial* and *institutional* conditions (such as the transport system, activities in place and institutional arrangements), *individual* conditions (personal needs and constraints) and *human interaction* which we want to achieve (diversity, intensity and duration).

Conclusions

The city under the socialist regime was understood as a physical site for containing industrial production. The city public space was often used for political 'rituals' and public manifestations, but it was not the basic unit for organizing consumption and did not constitute the means of production. The transition from state-organized collective to commodified consumption has brought profound changes within public space. New shopping malls, casinos, discos and bars are outcomes of the market economy and globalization trends.

If we take a look at the contemporary post-socialist cities we can see they are simply a collection of disconnected parts: luxurious shopping avenues and empty squares in the city centre, and neglected public space in peripheral housing estates. Connecting disconnected parts of the city through the means of public space and improving its quality in socially inclusive ways are the crucial factors for successful city spatial development in the future. In order to gain a better understanding of the possibilities of the diversity of places in the urban field, we should give more thoughts to the way in which spatial and social coherence is experienced.

The spatial model of the network city and the network thinking approach introduced in this chapter constitute steps towards the integration of different types of mobilities, experienced time and accessibility considerations into urban planning and design practice. By creating conditions for overlapping different 'landscapes' and by integrating the 'node quality' and 'place qualities' of a given location – encapsulated in the term 'accessibility' – public space can be seen as a system of places where diverse human interactions can unfold, and where virtual cities of different individuals might overlap.

The approach of the multinodal network creates possibilities to connect disconnected parts of cities (city centres, peripheral housing estates) and create the efficient network of public space. The forces that drive a city to function are generated by diversity and the need for information exchange between different types of these nodes.

In our opinion, urban planning and design strategies according to the network city approach, applied at the city level and local level in order to achieve conditions for spatial and social coherence, should be the leading concept that is most effective in influencing spatial developments in an increasingly mobile society. The preconditions for the future design policies should be rooted in the needs, constraints and perceptions of individuals and different social groups, where mobility can be seen as a source which generates these individual needs and constraints.

References

Aleksandrowicz. D. (1999). The Socialist City and its Transformation. *Discussion Papers*, 10. Frankfurt Institute for Transformation Studies, Online Library. www.europa-uni.de/de/forschung/institut/institut_fit/publikationen/discussion_papers/1999/99-10-Aleksandrowicz.pdf [accessed August 7, 2016].

Andrusz, G., Harloe, M. and Szelenyi, I. (1998) *Cities after Socialism: Urban and Regional Change and Conflict in Post-socialist Societies*. Oxford: Blackwell.

Bertolini, L. (1999). Spatial Development Patterns and Public Transport: The Application of an Analytical Model in the Netherlands. *Planning Practice & Research*, 14(2): 199–210.

Bertolini, L. and Dijst, M. (2003). Mobility Environments and Network Cities. *Journal of Urban Design*, 8(1): 27–43.

Boelens, L. (2000). *The Netherlands, Country of Networks. An Inventarisation of the New Conditions for Planning and Urbanism*. Rotterdam: NAi.

Burgess, R. (2007). *Technological Determinism and Urban Fragmentation: A Critical Analysis School of the Built Environment*. Oxford Brookes University, Online Library. http://ccs.ukzn.ac.za/files/burgess%20against%20technological%20determinism.pdf [accessed August 7, 2016].

Castells, M. (1996). *The Rise of the Network Society*. Oxford: Blackwell.

Davis, D. (ed) (2000). *The Consumer Revolution in Urban China*. Berkeley, CA: University of California Press.

Enyedi, G. (1998). *Social Change and Urban Restructuring in Central Europe*. Budapest: Akadémiai Kiadó.

Hajer, M. and Reijndorp, A. (2001). *In Search of New Public Domain*. Rotterdam: NAi.

Jacobs, M. (2000). *Multimodal Urban Structures: A Comparative Analysis and Strategies for Design*. PhD thesis. *Transformations*, 3. Delft: Faculty of Architecture, Spatial Planning Group, Delft University of Technology, Delft University Press.

Kostinskiy, G. (2001). Post-socialist Cities in Flux. In Paddison, R. (ed), *Handbook of Urban Studies*. London: Sage, 451–465.

Moiseeva, A. (2007). *Transit Space: Transformation of Public Space in Post-socialist Russia from an Urbanism of Network Point of View. Case Study: New Industrial Cities in Post-Soviet Russia*. Master thesis. Delft: Department of Urbanism, Faculty of Architecture, Delft University of Technology.

Petrovic, M. (2005). *Cities after Socialism as a Research Issue*. CsGG Discussion Papers. London: Centre for the Study of Global Governance.

Relph, E. (1976). *Place and Placelessness*. London: Pion.

Rooij, R. (2005). *The Mobile City: The Planning and Design of the Network City from a Mobility Point of View*. PhD thesis. Faculty of Architecture, Spatial Planning Group, Delft University of Technology, TRAIL Thesis Series T2005/1, the Netherlands TRAIL Research School.

Tuan, Y. (1977) *Space and Place: The Perspective of Experience*. Minneapolis, MN: University of Minnesota Press.

10 The restructuring of urban public space in the Baltic Pearl

Megan Dixon

Introduction

Since the fieldwork on which the following discussion is based took place – in St. Petersburg in fall 2006, March 2007, and July 2008 and in Beijing and Shanghai in January 2008 – events in Russia (e.g. on Bolotnaya Square in Moscow) and Ukraine (on the Maidan in Kiev) have focused attention on the critical opportunity (or not) to use public space for political communication. Trubina (2012) has commented perceptively that Russian demonstrators seemed to feel more confident about their protest if they gathered in a space regarded as more central, and thus presumably more efficacious. If these meaningfully central and thus highly symbolic spaces are barred to protesters, are their efforts in truth less effective? Here the attempts of St. Petersburg city authorities to block 2007 protesters from massing on Nevsky Avenue are also instructive. In the absence of viable 'use' of standard public space (standard in part because the state values it, too), what spaces work for creating civic identity? We must acknowledge important differences between gatherings that take place in key symbolic urban locations from the kind of 'public' that exists around otherwise private, domestic spaces; however, how these latter spaces function can tell us a lot about the role of the 'public' in a society.

This contribution to the volume explores concepts of public space through analyzing forms envisioned for public expanses in St. Petersburg, Russia, and in Beijing and Shanghai, China. The catalyst for the diffusion of 'public space' concepts from China to Russia – and for negotiations over differing concepts – is the Baltic Pearl, a 205-hectare development area southwest of St. Petersburg, financed by a consortium of real estate investment firms from Shanghai. Site observation and interviews with expert informants allow exploration of the envisioned content of 'public' space in a residential space, the Baltic Pearl development.

In analyzing the architectural forms of the spaces considered and the language used to describe them, I build on a Lefebvrian assumption that the potential for action in space depends in some measure on the spatial form in which the action occurs. Spatial form should not be understood to determine action, but it does provide significant constraints and possibilities. Use of Lefebvre's triad for spatial analysis allows us to trace the concepts and implied practices that shape public space and thus, in particular, to make clearer sense of the confrontation between differing notions of such space and of the implications for future practices in that space.

Lefebvre's triad of 'spatial practices, space of representation, and representation of space' allows analysis of multiple aspects of space. A useful illustration of the triad occurs in several places in *The Production of Space* when Lefebvre provides examples from medieval Europe (1991, 45). While the market square is a 'space of representation' where the society performs itself and its major assumptions, grounded by a cosmology or 'representation of space' that places the church at the centre of life and the earth nested in a series of invisible spheres, the 'spatial practices' in this example are the roads that connect peasant communities, monasteries, and castles and allow the important practice of pilgrimage. Analogously, we could frame the Baltic Pearl as a 'space of representation' for becoming European, enacting prosperity, and participating in the global economy. It reflects a 'representation of space' that sees the world as a network of 'world city' nodes that generate activity in surrounding expanses, shaped by a limited set of architectural visions of the good city. We can discern spatial practices in the Baltic Pearl through the types of activities envisioned as the focus of life (shopping instead of medieval pilgrimages) and the types of connections provided between residential and retail expanses.

Employing Lefebvrian analysis allows us to evaluate the potential for two common desires for public space: facilitating social or political debate and creating connections between citizens.

In terms of the first goal, the dominant Habermasian vision of public space extols it as the site of critically-reasoned public discourse, characterized by inclusiveness and universality; the loss of public space understood in this way poses a threat to open societies. Yet many have noted that the 'space of representation,' which Habermas envisioned, was the eighteenth-century coffeehouse (e.g. Mitchell 2003, 34), not the large squares that frequently claim to be 'public.' McLaughlin (2004) points out that scholars such as Holston have called for 'counterpublics' as an answer to the 'public' space that more frequently embodies state domination than free debate. Ultimately, we can distinguish between a 'public' space that expresses the consensus idea of the state and its symbols and a 'public' space that allows the expression of the 'will of the people,' i.e. political dissidence or debate. The same distinction is probably necessary in public spaces that are supposed to accomplish the second function of public space – fostering connections between radically different people. Sophie Watson calls for 'countersites' where alternatives to the dominant 'public' may take place (Watson 2006, 170).

Can the same material expanse have both state and 'people's' public functions? Certainly. However, positing or even creating a particular spatial form will not ensure the kind of activity envisioned when 'public space' arises in discussion (Mitchell 2003). Using Lefebvre's triad as a template suggests that we might evaluate public space for its true 'public-ness' (understood as providing opportunities for free assembly and critical exchange) and for its capacity to create connections by analyzing 'spatial practices' in particular, supported by consideration of 'spaces of representation' and 'representations of space.'

Crucially, in the following cases, the perceived need for 'public' space stays constant (a concept or 'representation of space') even as the forms proposed to satisfy it (the 'spaces of representation' and 'spatial practices') undergo change. In analyzing the shift from one historical period to another in the same nation-state expanse, Lefebvre asserted that 'the shift from one mode [of production] to another must entail the production of a new space' (1991, 46). In analyzing the case of the Baltic Pearl, the Chinese designers are producing a new expanse within a Russian culture of space. How will this affect the appearance and function of public space?

Deciphering the 'public' in the Baltic Pearl

While urban planning and architectural designs fit more neatly into Lefebvre's description of 'representations of space' (and have often been discussed as such; see Harvey 2001) the forms taken by designs for public space do suggest current spatial practices and future possible activities. In the case of the yet partly-built Baltic Pearl, analysis of actual activity is still impossible; the negotiations surrounding its form as designed, however, are rich with material for spatial analysis.

When the Shanghai investment consortium came to St. Petersburg in 2003 with the intention of building a multi-use residential-retail-recreational district, they wanted their project both to communicate understandably to the Petersburg administration and its population and also to express a vision of Chinese participation in shaping new global culture. Thus, there are elements in the design of the Baltic Pearl that express negotiation with the spatial-formal language of local Petersburg tradition and also elements that seem to express concepts of urban space employed in Shanghai and Beijing. To illustrate this, I will consider public squares and residential courtyards as examples of 'public space' in the Baltic Pearl.

The Baltic Pearl in the context of other urban projects

The Baltic Pearl project was announced and initiated shortly before a boom in high-profile urban projects related to the surge in energy-sector profits that Russia enjoyed beginning in 2004. These projects tended to involve prominent world architects and were cited often as elements increasing the prestige of St. Petersburg in the European and global context. Notable examples are the Mariinsky Theatre second stage (completed; originally designed by Dominique Perrault), New Holland (ongoing; designed by Norman Foster), the replacement for the Stalin-era Kirov Stadium (designed by Kisho Kurokawa), and the Okhta Center/Gazprom skyscraper (discontinued; designed by RMJM Associates).

The imperative to provide public space clearly resonated with developers. The highly controversial Okhta Center, for example, was intended to provide office space to several thousand employees of Gazprom Corporation subsidiaries; like the Baltic Pearl, it espoused a distinctly global aesthetic (although the chosen design negotiated much less with local Petersburg tradition). Before protests succeeded in ending the project, the corporation tried to soften the feared threat to the Petersburg skyline from the design's 300-meter height by emphasizing the provision of public or semi-public space in its advertisements: a full-page ad in the June 2008 issue of the tourist/expat glossy *Pulse* assured readers that the proposed skyscraper would include 'a generally accessible observation platform at 300 meters, cafés and restaurants for city residents, a year-round ice rink, the largest European museum of contemporary art, and linear parks in the tradition of Petersburg architecture' (*Pulse* June 2008, 19).

These promised amenities were recognizable as a language of global public space, deployed to reassure those who feared that the skyscraper would become a non-public eyesore, an opaque and menacing symbol of the corporate-cum-state power of Gazprom. They fit into 'the serial reproduction of "world trade centres" or of new cultural and entertainment centres, of waterfront development, of postmodern shopping malls, and the like' (Harvey 2001, 358), but their content was framed in partly local terms – not, apparently, as narrowly defined as the 'rather particular environment that is deemed appropriate to attract and keep the global elite' (Marshall 2003, 3) or delocalized, globalized 'premium networked spaces' (Brenner 2004, 244). However, while these claims may convince some, clearly the function of public space lies in the practices that become possible in it. As noted by a *New York Times* reporter covering the immanent opening of the much-heralded CCTV Tower in Beijing, designed by Rem Koolhaas, an architect may design a building using the formal language of public space only to have the new owner thwart his intentions:

> Long negotiations have unfolded over how much public access will be allowed: to the architect's distress, CCTV's directors have threatened to close off two public roads that cut through the site. An enormous plaza will also be restricted to the company's employees.
>
> (Ouroussoff 2008)

Those who opposed the Gazprom project feared that the same would be true in St. Petersburg.

In the rest of the chapter I will consider the discourse of public space compared to its implementation in material form in the Baltic Pearl and three comparative sites in China.

Discourse of public space in Baltic Pearl design editions

On paper the Baltic Pearl recognized the value of public space. In fact, in successive editions of its design plan books (August 2004, December 2004, December 2005, November 2006), language used to describe public areas revealed at least three approaches to their successful creation, usually involving residential courtyards.

First, the planners recognized public space as a means of mixing residents in social life. For example, in the August 2004 edition, after delegates from a Chinese design institute had studied the local conditions in St. Petersburg, they concluded that 'according to the investigation, the entrance [of the apartment building] is not only the access for each unit, but also the place for people assembly' (SIIC 2004a). The ideal of public space as 'accessible, rational, and suspending status hierarchies' (D'Arcus 2004, 357) also emerged in a quotation from a page describing 'neighbourhood centres,' which

> serve for daily life of residents. Through proper treatment of spatial scale and environment, a public space for different people in spite of class, income and age is formed, so communication between residents can be facilitated.
>
> (ibid.)

The designers thus caught at least the formalized discourse of communication in public space and deployed it 'as a strategy of distinction' (D'Arcus op. cit.). Second, in addition to 'communication,' planners promised to include security and spatial texture through incorporation of public space, as in this passage from the December 2004 edition which analyzed the use of courtyards (SIIC 2004b):

> Quiet and peaceful courtyards surrounded by residential buildings form a contrast with busy streets. Comfortable and tranquil courtyards make great contribution to safety of the community and communication between dwellers.

Third, if only briefly, the planners posited the kind of public space provided by court-yards as a mode of cultural integration. In the December 2005 edition, planners praised a design for the quarter submitted by a Swedish firm that

> successfully combines the cold regions of northern Europe and creates *a sunny courtyard* and *Chinese block-based urban life*. [. . .] It forms a clustered residence and a construction of public space, combining the experience of eastern Asia and northern Europe.
>
> (Tongji 2005, emphasis added)

This quotation indicates that the Chinese designers of the Baltic Pearl intended to focus on idealized public spaces that would facilitate social mixing. In implementation, however, more of a negotiation emerged.

Images of public space in the Baltic Pearl design

On one hand, the project expressed a traditional formal approach to the concept of the major public square. In the 2006 design, at the southern edge of the site, along the Peterhof Highway, a large structure with surrounding open space was depicted on the west side of the main entrance road (see Figure 10.1). The first publicly available bro-chure for the Baltic Pearl calls it the Southern Square. At its centre stood a tower in the form of a lotus blossom; this tower was not part of the general design negotiations for the project, but emerged from an independent architectural design competition held in Shanghai (Xu interview, 2008). The Russian architect working on the project in coop-eration with Chinese colleagues noted that this tower did not appeal to his supervisors or colleagues in the Petersburg planning system (Nikitin interview, 2006).

Figure 10.1 The Baltic Pearl: the Southern Square, St. Petersburg, Russia, 2008

In fact, Oleg Kharchenko, former chief architect of the city and now a member of the City Architectural Council, noted that the expanse surrounding the tower was overly large and forbidding, precisely the kind of space that he and his colleagues in planning and architecture wanted to avoid in new city projects. Kharchenko was unconvinced that the expanse would draw anyone to use it; he mainly perceived the resemblance to oversized Soviet-era squares, such as Moskovskaya Square at the origin of Lenin Avenue south of St. Petersburg, marked by a huge statue of Lenin. The tower seemed to him an example of anachronistic monumentality (Kharchenko interview, 2006).

By contrast, a Chinese professor of planning in a Shanghai university who was familiar with the tower design hypothesized that the tower had specific cultural significance.

> This is urban sculpture. For example, where there is an unknown soldier, there are many towers. This is a form to make something permanent. In Chinese culture understanding, this is the way of making something special. If we want to make someplace important, something grand, we make a tower.
>
> (Zhuo interview, 2008)

The four petal structures at the four corners of the square were intended to house specific functions: one was to be a recreational complex with cafés, restaurants, and movie theatre; one was to be a retail mall; one was to be a huge fitness complex; and the last was to be a hotel. As noted above, these are signature elements in new globalized 'public' spaces.

With such spaces, the real question is whether they can become usable and meaningful public space, allowing spontaneity and activities that give rise to connections between people. Moskovskaya Square with its Lenin statue is actually an intriguing example of such an expanse transformed. A nearby resident stated in an interview that the large square had been quiet and deserted for most of her time in St. Petersburg, starting in her student days. The addition of sunken fountains and numerous benches had allowed local residents to gather and sit, and this low-density usage eventually promoted additional activities such as skateboarding, bringing sizable crowds to the square on temperate afternoons and evenings. As designed, the Baltic Pearl's Southern Square seemed to have as its common practice the act of regarding the tower – a sign of the monumental space that Lefebvre deplored. Perhaps it is instructive that the end result of further designs was a grey block fronted by expansive parking, a neutral economic space without the pretense of globalized amenities (Dixon 2013a).

On the other hand, as of 2008 the overall project emphasized courtyards as transitional space between the clearly private area of apartments and the generally public area of streets and parks, and these have persisted in the actual construction. Early designs for the Baltic Pearl had small courtyards that seemed intended merely as functional exits from buildings. The presence of courtyards in the current design (evident in Figure 10.2) resulted from long negotiations with Russian colleagues over the form of the residential expanses. The Russian architect assigned to work with the Chinese, Sergei Nikitin, argued for the climatic and social value of these yards, as he described in an interview. Rather than have apartment buildings stand densely together all oriented in one direction, as he perceived them to be in Shanghai (in order to catch the maximum sun for drying laundry), he argued for the creation of microclimates and socially secure semi-public spaces.

Figure 10.2 The Baltic Pearl: residential courtyards, St. Petersburg, Russia, 2008

Yards have taken many forms in St. Petersburg, from the airless versions of the pre-revolutionary era to the moderately-sized Stalin-era yards to the imposing expanses between late-Soviet housing blocks. The popular conception of them is that they promote relations between neighbours, facilitating contact between residents of the same building as they enter the building and use the yard space for relaxation or supervising children. In order to help the Baltic Pearl become more 'Petersburg' in nature, Nikitin and his supervisors sought to include the courtyard form in the design (Dixon 2013b).

Certainly, a remaining question is whether the future inhabitants of the Baltic Pearl will be willing to act out this notion of public space with their neighbours. Mariusz Czepczyński (2008) has noted that in contemporary Poland the wealthy do not want apartment-block housing with courtyards; they want stand-alone villas with completely private expanses around them. The Baltic Pearl is definitely intended for the affluent class in St. Petersburg (*Nevastroyka* 2007). All of the efforts that went into creating the Baltic Pearl courtyards may be for naught if those who choose to live there do not share the classic Petersburg assumptions that yards promote neighborliness and public spirit. However, thus far at least three apartment complexes have proceeded according to this standard courtyard layout.

The contrast between the positive discourse on courtyards in Chinese-authored design books and the contested negotiation over Soviet-style squares and 'native' courtyards raises a series of questions. How are such contemporary projects designed and implemented in China, especially in cities where the investing firms in the consortium have been active? Do Chinese projects also value courtyards? If so, what shape do they take, and what 'public' functions occur there?

Models of 'global' public space from contemporary urban China

The Baltic Pearl design books rarely appealed to Chinese practices as a model for the project in St. Petersburg; the passage quoted above from October 2005 is a fleeting exception (although it testifies to a perceived commonality). However, there are clear analogies in Chinese urban forms, both traditional and contemporary. The courtyard houses of Beijing *hutongs* are a more distant model, as they were initially intended for the use of one family; however, aspects of life in Chinese yards and alleyways approximate the Russian idea, and the courtyard spaces around the four- to five-story apartment blocks built in the era of Sino-Soviet cooperation are similar. In Shanghai, too, there is an intense public life on the streets and sidewalks, while the numerous courtyards in colonial-era and colonial-influenced estate housing function similarly to the basic courtyard in Russia: they gather neighbours into contact with one another (see Figure 10.3, a small yard just off West Nanjing Road). The brief acknowledgement that there could be overlap in Chinese and Russian conceptions indicates a recognized commonality in the population's social needs.

Figure 10.3 A small yard just off West Nanjing Road, Shanghai, China, 2008

Satisfaction of the 'population's social needs' can be framed as a spatial practice that the design must fulfil. However, influenced by the concept of the Baltic Pearl as a 'global,' universal district, the 'space of representation' starts to shift. As noted above, the Gazprom advertisement in 2008 displayed a functional language of public space that recognized the need for social mixing, but instead of placing this mixing in a communal apartment or in a particular kind of residential courtyard, the designers of the project claimed that they could accomplish the same ends by including cafés and an ice rink.

Sharon Zukin (2008) justly noted that the *hutong* housing of Beijing and the *shikumen* and British estate housing of Shanghai (such as pictured in Figure 10.3.) are both being demolished at a rapid pace, thus perhaps demonstrating a desire among Chinese designers to avoid the courtyard as the relic of a discarded era. However, contemporary projects in China do reveal the courtyard form, and if the Russian courtyard vision is to be affected by the Chinese one, it is probably these ultra-modern courtyards that will influence the shift. The hypothesis is that these designs still value 'public' space, but have already reshaped the material form of the 'space of representation' for 'public' practice.

Observation of three completed projects in Shanghai and Beijing helps to understand what the rest of the Baltic Pearl might look like, and the kinds of 'public' spaces that will become available to the district's inhabitants and other residents from around the city. All three projects acknowledge the need for 'public' space in architectural form and accompanying promotional discourse.

Xintiandi

Xintiandi, shown in Figure 10.4, was a public-private project south of the upscale Huaihai Road. It replaced densely inhabited courtyard spaces such as the one in Figure 10.3. While the architects and developers were charged with preserving the architectural fabric (which they did to a moderate degree), the functions changed entirely. The two-block area now has shops, cafés, and upscale restaurants. Many tourists come here with cameras to capture this image of '1930s Shanghai.' A statement in the 'Xintiandi museum' asserts that Chinese people see Xintiandi as foreign, while foreigners see it as Chinese: it is meant to be a hybrid 'global' space, with a visible emphasis on youth, Westernization, commercialism, and cleanliness.

Notably, the space has more intimate dimensions than many shopping areas; it has retained its courtyard structure and allegedly promotes social mixing. Paul Agus, an architect involved in the project, claimed that 'when a girl has a new dress, she goes to Xintiandi' (Agus interview, 2008). On a 2008 visit, the area featured outdoor dining (a novelty for China at the time) and two coffeehouses (Starbucks and The Coffee Bean & Tea Leaf), which were both packed in late afternoon on two different days (January 2008). While a critic has called it 'the same finely wrought balance of theme park and shopping mall that increasingly passes for upscale urban life' (Iovine 2006), it clearly draws clientele.

When asked what districts in Shanghai provided a likely model for future sections of the Baltic Pearl, a firm representative suggested Xintiandi (Li interview, 2008). In fact, Xintiandi has been a popular model within China; similar projects have been completed in Hangzhou and Tianjin. It has also been named as the inspiration behind the controversial development by Beijing firm SOHO of Qianmen, the centuries-old

Figure 10.4 The Xintiandi project south of the upscale Huaihai Road, Shanghai, China, 2008

shopping street south of Tiananmen Square in Beijing that now houses Starbucks, Apple, and a series of upscale European retailers. The 'public' that it envisions is affluent and enjoys the urban entertainment of al fresco dining and human-scale spaces.

HaiShanghai

For comparison, a source in Shanghai (Zhuo interview, 2008) suggested a project near Tongji University completed by SIIC, the same parent firm that created the Baltic Pearl (the firm proudly mentions the Russian project on their website). HaiShanghai is nestled in a northern neighbourhood of the city, right next to standard mass housing but also near Tongji and other new developments. It has two black glass office buildings, an intimately-scaled commercial area, and tall apartment buildings that stand apart from the rest of the development (see Figure 10.5).

HaiShanghai displays innovative elements in its public space, such as whimsical public art (including a statue of a man taking a picture), diverse textures in the building décor, and artistic seating; a Starbucks café signals its participation in global public space.

The language describing HaiShanghai on the SIIC website (www.siic.com) calls it a Creative Commercial Street, 'an open space drawing endless imagination for creative persons.' The emphasis is on 'public' understood as business, global connections, and technical innovation. The intimate public spaces between the buildings, along with the theatre and lecture hall, are meant to prompt this creativity and the synergy between creative people that engenders yet more innovative activity. It is a public space designed in particular for the catalytic classes designated, for example, by Hannerz (1993), not necessarily

Figure 10.5 HaiShanghai creative street, Shanghai, China, 2008

for a broader mixing: the website also states that 'it leads the trend in Shanghai and gathers the people with great imagination and pursuits.' While the housing in this development is separate and thus does not participate in the courtyard-like spaces of the 'creative zone,' the smaller expanses for gathering people (at least briefly) – as they walk through, sit at a café, take a picture of the art, pause from work – are clearly a popular form.

Jianwai SOHO

For a final consideration of potential courtyard spaces and their concepts of 'public', Jianwai SOHO in the Central Business District of Beijing offers a provocative comparison. In this popular development just inside the Third Ring Road, housing towers surround central open areas with retail outlets and greenery; the courtyard effect is clearly visible in Figure 10.6. The insistent visual presence of advertising images and retail signs, typical of SOHO projects, places this district a world away from the intensely social-public spaces of *hutong* housing which used to exist not far from this spot.

Yet although the area seems to depart from an idealized public space as embodied in those traditional structures, creating dense contact between residents, it does have an explicit vision of mixing. The SOHO website claims that

> At the Jianwai SOHO Summer Carnival, which runs for four months each year, pop stars, poets, artists and writers present concerts, poetry recitals, street displays and theme salons to audiences who come in large numbers from all over the city.

Figure 10.6 The Jianwai SOHO development inside the Third Ring Road, Beijing, China, 2008

The developer clearly has an idea that the space needs to have a 'public' aspect; the content of this 'public' is cultural, seeing music, poetry, and literature as the uniting forum in which people make contact and form new connections. Thus, although the architectural transition in forms of 'public space' has forcibly evicted previous residents and their practices of public life, SOHO sees another 'public' taking shape: 'Jianwai SOHO has introduced not only a new style of housing, but also a new way of living.'

Again the Baltic Pearl

The Baltic Pearl uses similar language when it describes the contribution of the new district to St. Petersburg. The website (www.baltic-pearl.com) states: 'We want not simply to build a Euro-style universal district with developed infrastructure and European-quality services, but also to create a new way of life.'

What marks the projects that may influence the Baltic Pearl is an amenities-based concept of 'public' that also emerges in the Gazprom advertising. Whether proponents of a different, more political and social type of public space can affect the spread of this concept in the Russian or Chinese context is an urgent and open question. When I asked whether planners regretted the loss of the kinds of public space embedded in courtyards such as that in Figure 10.3, a Fudan University professor noted that concepts of public space are shifting in China itself:

The general trend is for social contact to take place not in the living area but in walking areas. Social relationships become less intense. But this is good! In the old courtyard style, it can be good but it can be bad, too. [. . .] In China now, people want more private space. [. . .] Green space in housing areas is used 80% by older people. Old persons want exercise. Young people want shopping malls: 'we are together, we see each other, we don't know each other, we don't talk to each other.'

(Zhuo interview, 2008)

Successful agitation for effective alternatives to the amenities-based concept of 'public' will require close observation of changing practices, concepts, and 'spaces of representation,' so that scholars' conception of what the 'public' wants keeps pace with what the 'public' uses.

'Global' public as a space of debate and connecting: an open question

Watson (2006) suggested that 'globalized space' does not facilitate the development of what we might think of as public in a broader political sense, since, as Marshall also notes, it tends to foster landscapes and service amenities that serve a particular homogeneous class of the global wealthy (2003). The 'global' spaces discussed here recognize the need to offer 'public' qualities (and this itself is intriguing), but it is not clear that they will offer the possibility of public debate or of true social mixing. D'Arcus asserts that 'it is in public space that difference is both displayed and encountered' (2004, 358). The presentation of a pre-conceived set of 'public' functions may limit their capacity to accommodate difference.

The Baltic Pearl project in Russia will be a test of these ideas. It may succeed in providing a common ground where affluent Russians and Chinese can mix and know each other better. Its very newness may disable the development of spontaneous social and political mobilization, since defense of local historical traditions tends to provide a platform for independence from intended state or corporate meanings (Scott 1998; for St. Petersburg, Boym 2001). The effect of Xintiandi and Xintiandi-influenced projects, such as Qianmen in Beijing, seems to be the hollowing out of local place-knowledge into simulacra of both tradition and public. As many have noted, an amenities-based concept of 'public' presumes the users of space to be consumers rather than citizens. But with time and the accumulation of local experience, an enriched 'public'-ness may provide material for both debate and connections between strangers.

References

Boym, S. (2001). *The Future of Nostalgia*. New York, NY: Basic Books.

Brenner, N. (2004). *New State Spaces: Urban Governance and the Rescaling of Statehood*. Oxford: Oxford University Press.

Czepczyński, M. (2008). Conference discussion *Public Space and Social Cohesion in the City: Present and Future*. St. Petersburg, July 3–4.

D'Arcus, B. (2004). Dissent, Public Space and the Politics of Citizenship: Riots and the 'Outside Agitator'. *Space and Polity*, 8: 355–370.

Dixon, M. (2013a). The Southern Square in the Baltic Pearl: Chinese Ambition and 'European' Architecture in St. Petersburg, Russia. *Nationalities Papers*, 41: 552–569.

Dixon, M. (2013b). Transformations of the Spatial Hegemony of the Courtyard in Post-Soviet St. Petersburg. *Urban Geography*, 34: 353–375.

Hannerz, U. (1993). The Cultural Role of World Cities. In Cohen, A. P. and Fukui, K. (eds), *Humanising the City? Social Contexts of Urban Life at the Turn of the Millennium*. Edinburgh: Edinburgh University Press, 67–84.

Harvey, D. (2001) [1989]. *Spaces of Capital: Towards a Critical Geography*. New York, NY: Routledge.

Interviews (2006). St Petersburg, Russia. November.

Interviews (2008). Shanghai, China. January 19–23.

Iovine, J. (2006). Our Man in Shanghai: Ben Wood Takes on History. *New York Times*. www.nytimes.com/2006/08/13/arts/design/13iovi.html [accessed August 13, 2006].

Lefebvre, H. (1991) [1974]. *The Production of Space*. Trans. by D. Nicholson-Smith. Oxford: Blackwell.

Marshall, R. (2003). *Emerging Urbanity: Global Urban Projects in the Asia Pacific Rim*. New York, NY: Spon Press.

McLaughlin, L. (2004). Feminism and the Political Economy of Transnational Public Space. *The Sociological Review*, 52: 156–175.

Mitchell, D. (2003). *The Right to the City: Social Justice and the Fight for Public Space*. London: Guilford Press.

Ouroussoff, N. (2008). In Changing Face of Beijing, A Look at the New China. *New York Times*. www.nytimes.com/2008/07/13/arts/design/13build.html [accessed July 12, 2008].

Pulse June 2008, 19. http://pulse-uk.org.uk/

Scott, J. (1998). *Seeing Like a State: How Certain Schemes to Improve the Human Condition Have Failed*. New Haven, CT: Yale University Press.

SIIC Shanghai Development and Investment Consultant Co., Ltd, and Vast United Enterprise (2004a). *'Baltic City' Concept Plan, St. Petersburg Russia*. August.

——— (2004b). *Baltic Pearl Urban Plan, St. Petersburg Russia*. December.

Nevastroyka (2007). The 'Baltic Pearl' Is Intended for the Petersburg Elite. March 17. www.nevastroyka.ru [accessed March 27, 2008].

Tongji University Planning and Development Institute (2005). *Joint Project 'Baltic Pearl' in St. Petersburg, Russia*. Shanghai, December. In Chinese and Russian.

Trubina, E. (2012). Protests and Cities: The Logic of Protest. *Russkii Zhurnal: Mirovaia povestka*. March 11. In Russian. www.russ.ru [accessed January 6, 2014].

Watson, S. (2006). *City Publics: The (Dis)Enchantments of Urban Encounters*. London: Routledge.

Zukin, S. (2008). Conference discussion *Public Space and Social Cohesion in the City: Present and Future*. St. Petersburg, July 3–4.

11 Public green space in Vienna between utopia and political strategy

Philipp Rode and Eva Schwab

Vienna's social democrat governments have a long tradition of taking utopian visions of social equity and well-being as a political reference for the development of the city. These rather abstract ideals have been challenged over time through place-based visions of urbanity and emancipatory processes inspired by them. The recent coalition between the Social Democrats and Greens represents – to some degree – the merging of concepts as different as that of a top-down welfare state and a bottom-up grass-roots initiative. In their working programme, public space is awarded an important role for urban development and urban life, and the way public green space in particular is conceptualised serves as an indicator for ideological negotiations regarding the development of urban society.

Drawing on these accounts, this chapter uses a critical planning approach to explore recent urban gardening practices in Vienna and trace the relationship between urban gardeners and planning politics at the interplay of utopian ideals, place-based disputes and applied planning policy.

Towards this end, we first outline the politics of urban green space through time from large- to small-scale programmes, from self-organization to Fordist provisioning. We highlight specific projects as manifestations or disputes involving hegemonic forces and then briefly describe the role public green space plays in the current coalition's programme and contextualise it with theoretical considerations regarding civil society in urban development. The second part of the chapter relates the findings of a survey on urban gardening practices in Vienna and highlights two diverging examples of the recent urban gardening trend.

The chapter ends with a discussion that brings together planning history and current practices to assess how local actors position themselves in the interplay of utopian ideals, place-based contestations and applied planning policy, and to highlight potential pitfalls.

Vienna: a growing city under red-green government

Vienna is the capital and largest city of Austria, with about 1.797 million inhabitants (MA 23 2015, 6). The city covers about 415 square kilometres, and is divided into 23 districts as political administrative units.

After decades of stagnation and decline in population, the city again recorded growth in 1990s. The population of Vienna has grown by about 10 per cent since the early 2000s (MA 23 2015, 7). This growth entails increased socio-spatial inequality, which manifests in territorial concentration of an ageing population and migrants' settlements, and is associated with a strained housing market (Rode et al. 2010a).

During the interwar period, Vienna was shaped by socialist mayors, and has been governed by these since 1945. Only twice have they failed in elections to obtain an absolute majority of seats: they entered into a coalition with the Conservative Party from 1996 to 2001, and have been in a coalition with the Greens since 2010. The political and economic stability contributes to a comparatively late manifestation of international trends in Vienna, such as governance changes, a neoliberal orientation of the local state and global competitiveness (Dangschat and Hamedinger 2009, 95f).

The proportion of green space in the entire city is approximately 45.5 per cent, wherein the largest share (almost 87 per cent) consists of extensively used and designed agricultural and forestry land (www.wien.gv.at/statistik/lebensraum/tabellen/gruen flaechen-bez.html, accessed March 16, 2014). It is located in the forest and meadow belt, which was established in 1905 to preserve the landscape at the outskirts of the city for recreation and conservation. Accordingly, the proportion of green space in each district varies between 3 and 13 per cent in the densely built-up inner city districts to 70 per cent in the western outskirts (ibid.).

Since the mid-2000s the civil society has shown growing interest in urban gardening, representing bottom-up processes capable not only of revising planning approaches to urban green space but also of changing urban society. On one hand, this development makes reference to contestation discourses such as 'the right to the city', 'the Commons' or 'guerrilla gardening'. On the other hand, the city government reads and employs this trend as a contribution to the high quality of life in Vienna, which represents an essential location attribute for the Austrian capital in the international competition between cities (www.wien.gv.at/politik/international/wettbewerb/mercerstudie.html and www. wien.gv.at/politik/international/wettbewerb/rankings.html, accessed April 5, 2014).

A partial history of Viennese planning politics

A focus on large-scale projects in the field of housing and infrastructure is characteristic of the town planning policies of socialist governments since the early twentieth century. However, between 1918 and 1924, a cooperative self-help movement influenced both the settlement movement as well as the formation of allotment garden associations. The notion of self-sufficiency through production of fruits, vegetables and small domestic animals in one's own garden was as important as the joint construction of residential buildings and community facilities.[1] The socialist settlement cooperatives tried to establish a project for societal change with their activities, encompassing an educational mission with local self-administration in small networks and socio-economic reforming approaches (ibid., 149 and 154).

The settlement movement came to an end in 1924, when the socialists opted for large-scale social housing, now known as the 'Superblock' of 'Red Vienna'. Ideals of joint self-help and self-management had to yield to an all-providing local state. The emancipatory moment of the cooperatives was lost by the subsequent generations in favour of a highly regulated possessive individualism (ibid., 157).

After years of reconstruction in the aftermath of World War II, large-scale projects were revived in the 1960s, when the dense, inner districts of the 'Gründerzeit' lost much of their residential population, who moved to the large housing estates in the peripheral districts. The overall large-scale planning strategy was also reflected in the green space policy, as shown in peripheral developments such as the Donaupark

(1960–1964, 60 hectares), the Kurpark Oberlaa (1969–1974, 86 hectares) and the Donauinsel (1972–1988, 3.9 square kilometres).

Addressing the problems of the inner city, the Fordist approach was expanded in the 1970s to include smaller projects of 'urban renewal' (Mattl 2012, 27). The Sternwartepark, a small park located in an upper-class cottage area, played an important role in this paradigm shift: it was the then-socialist mayor's aim to use part of the park for residential construction. A local citizens' initiative was formed against the project, arguing for the preservation of the park and its trees. The initiative soon received support from the opposition parties as well as a major daily newspaper, converting the local significance of the park into a city-wide issue, for which the first Viennese referendum was held. It ended with a majority of 57.4 per cent for the preservation of the park. The acting mayor resigned after this defeat (Bihl 2006, 616).

Subsequently, the model of the multi-functional and mixed city was partially integrated into the principles for spatial planning (Meissl 2006, 698), and urban renewal offices were established. Up to now, these have mediated between the administration, land owners and users in small-scale projects. At the same time, autonomous activists self-organised beyond this new institutional framework. By squatting in abandoned spaces, the movement developed into a 'resistance of places' (Mattl 2012, 28). The 'Burggarten' movement, as a part of this initiative, explicitly addressed rigid utilization standards in the city's central green spaces and in the late 1970s used sit-ins to challenge these (Wiener 2012, 146). This dispute led to a redefinition of publicness in central and historical locations of the city, bringing the issues of representative function vs. recreational usage into discussion.

These areas of conflict since the mid-1970s reflect the growing self-awareness of civil society as well as an institutional adaptation. Since the 1990s, a change in town planning towards the entrepreneurial city has been observed (Novy et al. 2001, Lička et al. 2013). The first of the projects aiming at strengthening Vienna's position in the competition among cities, such as the EXPO, highlighted the contradictory planning approaches of the Social Democrats and the Greens. While the Social Democrats continued to focus on large-scale projects, the then oppositional Greens targeted decentralized, community-based, small-scale undertakings (Cattacin 1994, 121).

Since then, the role of the local state has been formally reduced, as large-scale projects have been initiated by development companies with close links to the municipal government. Simultaneously, the influence of democratic bodies has been reduced in the decision-making structures of governmental development agencies and companies (Novy et al. 2001, Seiß 2007).

Viennese city planning has increased its investment in public spaces since the early 2000s, and established them as a versatile domain of politics. Public space plays an essential role as a medium of change for the renovation and development of the city. Thus the issues of social integration, urban development and environmental protection have been linked with the development, design and use of public space, especially in local, small-scale projects. This dynamic has increased in the last ten years (Knierbein et al. 2014).

Concentrated intervention in places frequented by tourists and having economic significance can be observed. This contrasts with political and planning objectives and shows that the multiple expectations regarding public spaces are not always reconcilable (ibid.).

Public green space in the red-green government coalition

Since 2010, Vienna has been governed by a coalition of Social Democrats and Greens – parties with contradicting approaches to city planning until now. The planning department is headed by Greens. The department for the environment, which administrates urban green spaces, is led by a Social-Democratic councillor. An approximation of city planning politics can be identified, especially in the joint approach to open public spaces, in the government agreement between the parties. There is an emphasis on 'attractive' design of public space as a significant factor in the urban quality of life. The government agreement addresses the involvement of civil society at the local level, with pilot projects playing a major role (Häupl and Vassilakou 2010, 58f). Self-harvesting and community gardens are explicitly mentioned as welcome (ibid., 59).

At the macro level, the red-green coalition primarily recognises green space as recreational and ecological compensation. Furthermore, an impetus for change is addressed: even though the emancipatory content of the pilot projects is not explicitly mentioned, the local level and socio-productive uses of public green space are addressed as a testing ground for alternative lifestyles and city models. It seems as if a place-based and bottom-up approach has been pushed forward, presenting green space as a medium to develop an urban utopia on a small scale. The focus lies on a contribution to social and cultural change, not to technical or design change – as architectural and urban planning utopias usually tend to do. With the parties' diverging conceptions regarding city planning and urban development in mind, it is worth taking a look at how the utopian possibilities of small-scale approaches to urban development are reflected in theoretical discourses.

Civil society initiatives: between utopia and privatization

The formulation of utopias overlaps with crises of accumulation regimes and, particularly, civil society initiatives which test new models of society in the current neoliberal crises (Laimer 2013, 6). Utopias are often dismissed as ivory-towerish during periods of (relative) stability. The dilemma of a utopia is intrinsically rooted in its ideas, which significantly differ from the criticised structure and thereby appear unrealistic. Fainstein (2010) points out that utopias and revolutionary acts do have important theoretical content, but are difficult to implement in practice and often lack the support of the majority of society. Fraser (2003) addressed this problem by postulating 'non-reformist reforms', which gradually become more radical with every step and therefore have the power to trigger change within an existing social framework. To mediate between vision and reality, Lefebvre (1966) introduced the term 'autogestion', focusing on routine articulations and practices of utopia by addressing the tools of participation, appropriation and self-empowerment.

Small-scale interventions in public and green space (cf. Mörtenböck and Mooshammer 2013, Reynolds 2010) aim to change its meaning and bring social change, thus fitting Castells' definition of urban movements. The minor, everyday initiatives and associations are the ones that develop emancipatory content within existing political structures and initiate social and cultural change (Amin and Thrift 2004, 234). Particularly, means of sensing the city – as in urban gardening, for example – are ascribed the power to become 'weapons of emancipation and political struggle' (ibid., 233). The aesthetics of informality or organisational traits of community gardens are connected to this (ibid., 234). Another source for forms of emancipatory urbanism can be found in the theory of the Commons (www.blog.commons.at/commons/, accessed March 11, 2014). Marcuse (2009) amends its local focus by defining Commons Planning as a comprehensive approach focused on common welfare in urban developments. The supra-local perspective of emancipatory urbanism comprehends the interplay between local initiatives and global discourses as a potential renewal for forms of contestation. From a global perspective, the wealthy cities have become a type of 'global suburbs' and 'constitute privileged spaces for more or less gentrified "creative classes", and breeding grounds for a mix of alternative, critical, and "bohemian" milieus' (Mayer and Boudreau 2012, 278). These environments give rise to urban movements, which take on the challenges to corporate urban development, calls for social and environmental justice, and anti-globalization (ibid., 284). An essential form of their expression are 'small acts of reappropriation of urban space' (ibid., 285). Mayer and Boudreau argue that this approach is found mainly in dense cities and dense social networks, i.e. within a circle of existing acquaintances, and that the well-known forms of expressions are urban gardening, interventions in public (road) space and activities in public parks (ibid., 286).

Harvey critically points out a 'militant particularism' (1996, cit. in Mayer and Boudreau 2012, 276) with conservative and self-referential motives within these local and spontaneous initiatives. His claim for a 'politics of solidarity' takes the global context into account, just like Marcuse postulates for Commons Planning.

As has been shown, small-scale projects in public green space may be attributed a significant role in the gradual and small-scale change of urban society. However, in Vienna a structural and actor-related analysis of the initiatives, enabling their precise discursive positioning, as well as a discussion of resulting administrative-political conflicts in terms of utopian change, are missing. In this context, it is of interest how the notion of urban green space has expanded to emancipatory aspects. Therefore, the current forms of urban gardening are represented in an overview and thereafter discussed in detail with the help of two case studies.

Forms of urban gardening in Vienna

Parallel to the emergence of public space as a policy and means of distinction in the inter-city competition of the 2000s, temporary (art) initiatives have made the issue of public space more accessible and initiated a debate about its qualities by introducing terms such as accessibility, justice, and social and environmental sustainability (Rode et al. 2010b). The beginnings of certain urban gardening projects were part of these artistic undertakings, as the case study on Heigerleingarten will show. The first projects were documented in 1998 and 2001. Intensification in the initiation of garden projects since 2009 can be identified; it reached its height in the years 2011 and 2013, with ten new initiatives respectively.

A survey conducted in spring 2014 shows that about 45 gardens and green spaces in the city of Vienna are used as community gardens. A large part of these projects (about 58 per cent) are situated in densely built-up, historic urban space. A further 13 per cent are located in urban development areas or in newly constructed projects, and 11 per cent in residential neighbourhoods at the outskirts of the city. About 7 per cent of the projects are found in large-scale settlements of the Fordist type.

About half of the projects are located in existing parks; more than one-third of these show a particular spatial configuration (such as an internal courtyard, linear layout or micro open-space) or utilize underused border areas. Approximately 13 per cent of the projects use open spaces of residential developments. Another 9 per cent are active in internal courtyards or portions of built-up plots. Approximately 11 per cent are located on abandoned or underused land; only 7 per cent of the projects are located on agricultural land. The ownership structure mirrors these findings: almost 70 per cent of the land is owned by public authorities. In 9 per cent of cases, developers or investors are the owners of the land, and private individuals own about another 9 per cent.

A long-term intention can be distinguished for most projects; only somewhat more than 13 per cent of the projects are explicitly designed as temporary. In almost one-third of the cases, non-profit organisations (NPOs) in the immediate vicinity initiated the projects. Among the NPOs, the Gartenpolylog association plays a major role, which can be described as a network node; it initiates nearly 20 per cent of all projects. In more than 13 per cent of cases, Local Agenda 21 acted as the trigger of a project; the urban renewal offices and private-sector actors initiated another 9 per cent of projects each; 28 per cent of projects are driven by citizens and local residents.

The analysis of the mission statements of the various projects – found in the associations' bylaws, on websites and interviews in the media – shows that the majority of the projects (about two-thirds) are organized as an association. Only in two cases is there no formal management body of the project, and the actors remain anonymous. In all cases, the project was focused on enhancing the neighbourhood at the local level. In a few cases, a more individualistic focus on improving mental and physical well-being and promoting gardening as a hobby became apparent. Other central issues were environmental education, ecologically oriented concerns – such as self-supply – and the aspect of democratic self-organization for social change. A utopian streak can be observed in 13 per cent of the projects, e.g. the development of urban alternatives is a target, and only 4 per cent of the projects aimed at formulating protest.

In order to show the scope of the projects regarding content and structure, two initiatives that strongly differ in their organisational form, their intention and their spatial implementation are discussed in detail.

Intercultural garden

Heigerleingarten was launched in 2008 as the first neighbourhood garden of its kind in the city (www.gartenpolylog.org/de/3/wien/16.-bezirk/nachbarschaftsgarten-heigerlein/praesentation, accessed February 8, 2014). The project had a temporary forerunner as part of a public art festival and followed the example of 'intercultural gardens' in German cities. The implementation of the festival garden served as a test run, during which acceptance problems at the institutional and neighbourhood level were resolved (ibid.). A negotiation process with the city government and administration accompanied the establishment of Heigerleingarten, the final result of which was a new policy for dealing with future urban garden initiatives.

The founders of the Gartenpolylog association were primarily women; they can be described as young, educated and professionally experienced in community gardening, sustainability and interculturalism as well as in public relations and teaching activities (www.gartenpolylog.org/de/1/uber-uns, accessed February 12, 2014).

The Heigerleingarten (approximately 1,200 square metres) is a fenced part of a public green space, and has a cultivated area of around 200 square metres in raised beds. It is located outside the city centre in Ottakring, a multi-cultural working-class district with some gentrified 'bohemian' hotspots, along a railway line. The surrounding green space was previously unused, as mentioned by the City of Vienna (https://www.wien2025.at/site/urban-farming-gartnern-in-der-stadt/, accessed March 12, 2014). The garden is visible, but is separated from the surrounding green space by a one-metre-high steel fence and is accessible only to members. The association consists of 50 members, and 15 of them have keys to the garden. These members maintain the raised beds for their private use (Møltoft Jensen 2013). The garden merges the interests of existing local institutions and of local residents and provides them with a new field of action in this 'natural space of experience' (www.gartenpolylog.org/de/3/wien/16-bezirk/nach barschaftsgarten-heigerlein/praesentation, accessed February 8, 2014). Meanwhile the gardeners of the Heigerleingarten have established their own association, which is a contractual partner of the City of Vienna based on a licence agreement. They grow mainly herbs and vegetables.

A number of initiatives followed the example of Heigerleingarten, partly with the assistance of Gartenpolylog, taking up the city's then-new offer to support neighbourhood gardens.

Guerrilla garden

Längenfeldgarten has evolved from a guerrilla gardening initiative launched in 2010.

Since 2009, the members of the decentralized network for art, culture and media alternatives (Kukuma) have been active in terms of guerrilla gardening and have gardened at several locations throughout Vienna (Pöltner-Roth and Kromp 2013, 31). In all locations except Längenfeldgarten they have experienced conflicts with the landowners.

The initiative proposes to create self-made, liveable urban spaces through self-determined action. They explicitly disapprove of property and real estate speculation, and advocate collective spaces as 'Space for Utopia' (www.kukuma.org/, accessed February 26, 2014).

While the political and social context of guerrilla gardening and the members' motives are published on Kukuma's website, the people involved remain anonymous. It is specified that 25 people are currently active in the garden. Gardening by other people is explicitly encouraged.

The area for vegetable beds, of approximately 400 square metres, has been utilized without any consultation or agreement with the owner, the Wiener Linien – Vienna's public transport organisation. Since occupying the area in 2010, Kukuma has repeatedly informed the owners about its intentions but received no response. Thus, it can be argued, the initiative is tolerated by the owners. Although the users indicated that they had experienced vandalism while initiating the gardens, once they have been established they are no longer the targets of attacks (Møltoft Jensen 2013).

The beds are situated in a hardly visible part of a public green space, surrounded by underground tracks. The cultivable land can be accessed by anybody. Recycling materials are used for contouring the irregular patches. There is an area for storing garden tools as well as compost piles (Møltoft Jensen 2013).

Gartenpolylog and Kukuma manifest two contrasting forms of organisation: Gartenpolylog has developed from a local association into a city-wide service provider and initiator of community gardens, and thereby became a partner of the city. Kukuma has positioned itself in opposition to legitimate forms of urban gardening by deliberately rejecting hierarchical structures and ownership ideas. While the Kukuma case can be read as encouraging in terms of utopian content and political messages, it is worth noting that the City of Vienna has attempted to de-politicise the guerrilla garden discourse (https://www.wien2025.at/site/urban-farming-gartnern-in-der-stadt/, accessed March 12, 2014).

The local level is the starting point of initiatives, both spatially and socially. Neighbourhood initiatives and small networks of like-minded people initiate the interventions. Inspired by international examples, they explore political, social and ecological areas of action. The social aspect has been proven to be an essential factor in the emergence of an initiative as well as in the durability of their commitment. In the current urban gardening initiatives, initiators consist of existing groups of people who link their private interests with an expression of values and social commitment and mainly use the support offered by the City of Vienna. The synopsis of the Vienna initiatives reveals that the actors are to be found in dense networks of educated and creative-class milieus which form the 'global suburb'.

Institutionalization vs. utopia?

Two essential characteristics become manifest through the analysis: the Viennese urban gardening scene was formed relatively late and is strongly institutionalized. The first aspect is associated with the above-mentioned slow manifestation of global trends in Vienna. Socio-economic indicators like accelerating segregation can be interpreted as a symptom of a crisis which occured with a delay in comparison with other European cities. If the emergence of urban gardening is seen in connection with these indicators of crisis only, then some utopian content could be assumed.

However, the second aspect introduces another interpretation: while the start of most initiatives can be traced back to civil society engagement, the structural and spatial implementation shows attributes of a settled movement, integrated in and regulated by structures of the local state and its institutional network.

This is due to the communal policy which ties the allocation of public land to the establishment of certain structural bodies for taking over the responsibility for the utilization of plots. Therefore, the initiatives are long-lived and have committed gardeners. However, they can hardly be called an alternative to corporate urban development. This raises the assumption that such a strong integration of social movements into enduring bureaucratic structures weakens their utopian content.

This can be ascertained from the historical reconstruction of the settlement movement, as shown in the first part of the chapter. Translated to today's urban gardening movements, they run the risk of becoming a fig leaf of urban development through institutionalisation and standardisation.

An extended view on urban green space

The initiatives' focus on public green space can be interpreted as a sign of dissatisfaction concerning the administration of green spaces, and criticism of the traditional understanding of urban green space. The intentions of the initiatives are clearly connected to self-empowerment and self-organisation. A main achievement has been a conceptual and discursive expansion of the notion of urban green space involving aspects of production. Thus urban green space is no longer reduced to compensation, representation and recreation opposing the urban. In contrast, green space is considered an integral part of the city by the urban gardeners, who see themselves as producers.

This conceptual expansion involves negotiation processes that range from direct resistance to compromises in social partnership style. The Burggarten case clearly reflects the contrast between civil society and administrative bodies regarding their ideas about the use of public green space. Back in the 1980s, the administrative bodies used an authoritarian interpretation of their role, which came into direct conflict with the protesting youths. Nowadays the structures of civil society appear to be established enough for the initiatives to enter the negotiation process. This process can be described as a succession of certain actors, with art and culture initiatives as a pioneering force. The global suburbs follow suit, organising themselves in dense networks and being capable of negotiating and articulating their needs. However, they cannot be described as homogenous, as their backgrounds span from utopic leftist to urban professionals with commercial interests, united in a hedonistic flair. This diversity of the gardening crowd reflects the wide variety of meanings which green space is capable of conveying. Moreover, the gardening initiatives represent a valorisation of spaces which is inexpensive to the municipality, but contributes positively to the image of the city and further establishes public space as a place for social and ecological sustainability.

Social re-densification and planning shortfalls

From a planning perspective the emergence of initiatives reflects two different tendencies.

Firstly, the re-densification of already densely built-up areas is also visible in open space – micro open spaces that were designed as interspaces or flower beds are socially densified by active use; thus they are replaced by a more user- and productivity-oriented design. This is again linked to the process of institutionalisation, as communally funded gardens tend to have standardized design. Improvised or self-made designs, which would extend the communal notion of design, constitute the exception.

Secondly, the gardening initiatives' use of existing parks and green spaces can be interpreted as a critique of their design and role in urban development, since open spaces with difficult formal configurations and fragmented layouts are a clear sign of their subordinate role as urban leftovers. Unfavourable configurations of linear or fragmented open spaces around the inner urban fringe have received new use and significance through the initiatives.

Possibilities and pitfalls

The Viennese urban gardening initiatives cause the renewal of urban green space on different levels. The broad concept of gardening is ideally suited for being charged with many different desires – ranging from philistine, individualist allotments to emancipatory spaces to marketing tools. The majority of the Viennese garden initiatives focused on socio-cultural strengthening of local neighbourhoods and social cohesion within distinct communities. The smaller part of the initiatives deals with the development and implementation of urban alternatives critical of corporate urban planning. These more utopian initiatives also tend to have a more emancipatory content, which manifests itself in a stronger autonomy in structure and implementation. The predominant focus on the local level, however, might hinder activating emancipatory possibilities. In the absence of exchanging and networking, individual comfort concerns could replace critical views. Solidarity and emancipation require constant exchange and development, as already seen in the early garden movement in Vienna. Another pitfall might lie in observable tendencies of privatization of public space. Fencing off can be called a distinguishing garden feature, but this socio-spatial practice applied in public space reduces its accessibility and publicity. While this may be due to vandalism experiences, it also reflects particular interests of the members. These aspects exemplify Harvey's criticism and suggest that the notion of Commons Planning could present an important expansion of the local-level initiatives fostering their emancipatory power.

The attitude of the current government is to be characterized as ambivalent. While the support of the initiatives should be evaluated positively and also reflects the willingness to further develop the concept of urban green space, the phenomena of design standardisation and structural institutionalisation represent clear signs of control, which also carry streaks of discursive misappropriation. If these tendencies continue, there is a risk that the initiatives become ever more self-referential. The small-scale approach of urban gardening has become to some degree a mainstreamed strategy, which is tactically exploited for marketing issues in commercial and political aspects. The Viennese policies contain the risk of choking off alternative expressions of the urban gardening idea and thus hindering the constant revival, critique and contestation that are the foundation of utopias. Propelled by utopian ideas, the urban gardening initiatives have undeniably brought about renewed concepts of urban green space and strengthening of civil society initiatives. The observed institutionalisation and particularism, however, reveal a problem: lacking emancipatory content, the urban gardening hype is bound to produce privatised or underused urban spaces awaiting the next crisis to be laden with utopian desires.

Note

1 Access to private residential buildings for blue-collar workers was provided only by cooperative organisations (Novy 2012, 135).

References

Amin, A. and Thrift, N. (2004). The 'Emancipatory' City? In Lees, L. (ed), *The Emancipatory City: Paradoxes and Possibilities*. Thousand Oaks, CA: Sage.

Bihl, G. (2006). Wien 1945–2005 – eine politische Geschichte. In Csendes, P. and Opll, F. (eds), *Wien – Geschichte einer Stadt Band 3*. Vienna, Cologne and Weimar: Böhlau Verlag, 545–650.

Cattacin, S. (1994). *Stadtentwicklungspolitik zwischen Demokratie und Komplexität – zur politischen Organisation der Stadtentwicklung, Florenz, Wien und Zürich Vergleich*. Frankfurt/Main and New York, NY: Campus Verlag.

Dangschat, J. and Hamedinger, A. (2009). Planning Culture in Austria – The Case of Vienna, the Unlike City. In Knieling, J. and Orthengrafen, F. (eds), *Planning Cultures in Europe. Decoding Cultural Phenomena in Urban and Regional Planning*. London: Ashgate, 95–112.

Fainstein, S. (2010). *The Just City. New York, NY*: Cornell University Press.

Fraser, N. and Honneth, A. (2003). *Recognition or Redistribution*. London: Verso.

Häupl, M. and Vassilakou, M. (2010). *Gemeinsame Wege für Wien – Das rot grüne Regierungsübereinkommen*. www.wien.gv.at/politik/strategien-konzepte/regierungsueberein kommen-2010/pdf/regierungsuebereinkommen-2010.pdf [accessed January 17, 2014].

Knierbein, S., Madanipur. A. and Degros, A. (2014). Vienna: (Re)framing Public Policies, (Re) shaping Public Spaces? In Madanipur, A., Knierbein, S. and Degros, A. (eds) *Public Space and the Challenges of Urban Transformation in Europe*. London: Routledge, 23–37.

Laimer, C. (2013). Das urbane Leben hat noch gar nicht begonnen. *Dérive – Zeitschrift für Stadtforschung*, 53: 4–8.

Lefebvre, H. (1966). Theoretical Problems of Autogestion. In Brenner, N. and Elden, S. (eds), *State Space World – Selected Essays Henri Lefebvre*. Minneapolis, MN: University of Minnesota Press, 138–152.

Lička, L., Rode, P. and Bistricky, D. (2013). Open Space for Social Housing – between Social Benefit and Marketing Asset? In Schrenk, M., Popovich, V., Zeile, P. and Elisei, P. (eds), *Planning Times*. Proceedings. Rome: Real Corp, 661–670.

MA 23 (2015). *Wien in Zahlen*. Vienna: MA 23.

Marcuse, P. (2009). From Justice Planning to Commons Planning. In Marcuse, P., Connolly, J., Novy, J., Olivo, I., Potter, C. and Steil, J. (eds), *Searching for the Just City: Debates in Urban Theory and Practice*. New York, NY, and London: Routledge.

Mayer, M. and Boudreau, J-A. (2012). Social Movements in Urban Politics: Trends in Research and Practice. In Mossberger, K., Clarke, S. E. and John, P. (eds), *Oxford Handbook on Urban Politics*. Oxford: Oxford University Press, 208–224.

Mattl, S. (2012). Der Mehrwert der urbanen Revolte. Die Erneuerung Wiens aus dem Geist der Hausbesetzer. In Nußbaumer, M. and Schwarz, W. M. (eds), *Besetzt! Kampf um Freiräume seit den 70ern*. Vienna: Czernin Verlag, 22–28.

Meißl, G. (2006). Ökonomie und Urbanität. Zur wirtschafts- und sozialgeschichtlichen Entwicklung Wiens im 20. Jahrhundert und zu Beginn des 21. Jahrhunderts. In Csendes, P. and Opll, F. (eds), *Wien – Geschichte einer Stadt Band 3*. Vienna, Cologne and Weimar: Böhlau Verlag, 651–738.

Møltoft Jensen, M. (2013). Allotment and Urban Gardens in Vienna. From the Inner City to the Urban Fringe. www.urbanallotments.eu/fileadmin/uag/media/STSM/Allotment_and_urban_gardens_in_Vienna_-_Part_1.pdf [accessed January 13, 2014].

Mörtenböck, P. and Mooshammer, H. (2013). Platzbewegungen – Expeditionen in den Raum der Versammlung. In Lange, B., Prasenc, G. and Saiko, H. *Ortsentwürfe – Urbanität im 21. Jahrhundert*. Berlin: Jovis, 92–100.

Novy, A., Redak, V., Jäger, J. and Hamedinger, A. (2001). The End of Red Vienna: Recent Ruptures and Continuities in Urban Governance. *European Urban and Regional Studies*, 8(2): 131–144.

Novy, K. (2012). Selbsthilfe als Reformbewegung – der Kampf der Wiener Siedler nach dem 1. Weltkrieg. In Krasny, E. (ed), *Vom Recht auf Grün*. Vienna and Berlin: Turia and Kant, 126–159.

Pöltner-Roth, K. and Kromp, B. (2013). *Miteinander Garteln in Wien. Im Auftrag der MA 49*. https://www.wien.gv.at/umwelt-klimaschutz/garteln-bilanz.html [accessed April 4, 2014].

Reynolds, R. (2010). *Guerilla Gardening – ein botanisches Manifest*. Freiburg: Orange Press.

Rode, P., Giffinger, R. and Reinprecht, C. (2010a). Soziale Veränderungsprozesse im Stadtraum. *Werkstattberichte Nr. 104*. Vienna: Magistrat Wien.

Rode, P., Wanschura, B. and Kubesch, C. (2010b). *Kunst macht Stadt – Vier Fallstudien zur Interaktion von Kunst und Stadtquartier*. Wiesbaden: VS Research.

Seiß, R. (2007). *Wer baut Wien? – Hintergründe und Motive der Stadtentwicklung Wiens seit 1989*. Salzburg and Munich: Pust.

Wiener, S. (2012). Erinnerungen an die Anfänge der Burgarten-Bewegung. In Nußbaumer, M. and Schwarz, W. M. (eds), *Besetzt! Kampf um Freiräume seit den 70ern*. Vienna: Czernin Verlag.

12 The normative construction of a (public) urban space through the use of policy instruments

Some reflections from northern Italy

Michela Semprebon

Introduction

Studies on policy instruments have been crucial to understanding the late twentieth century transformations of the state, with the resulting processes of public sector reform and devolution (Kassim and Le Galès 2010). Various strands of analysis have revived classic questions in relation to whether, when, what and how governments govern and interact with citizens. Research has stressed, for example, the deceptive apparent neutrality of instruments and how it has been used to create consensus on policy initiatives while hiding implicit political goals (John 2011; Kassim and Le Galès 2010).

The hypothesis of this chapter is that the adoption of certain spatial policy instruments can be associated with a trend of increasing politicisation of urban safety issues which has characterised the last two decades, whereby innovative policy inputs have been frustrated by efforts to 're-order' (public) space in response to voters' anxieties. In order to address this question, empirical evidence will be borrowed from the legislative process that led to the definition of an apparatus to regulate phone centre shops in Verona. From the theoretical point of view, the chapter engages with the political sociology approach to policy instruments (Lascoumes and Le Galès 2004; Halpern et al. 2014) to investigate the link between instruments themselves and the mechanisms and dynamics of public action, at different scales, with attention paid to both the dimension of policies and politics.

The main scope of the contribution is to provide an example of the interweaving political dynamics entrenched in legislative instruments. It aims to contribute to discussions on urban politics and democracy while showing how these dynamics have worked against the development of innovative forms of interventions on phone centre shops that could in fact foster social inclusion.

The first section grounds the chapter in the theoretical debate, while providing a short punctual overview of the relevant literature. The second introduces the methodology and the case study. The third focuses on the empirical findings to draw some final reflections and conclusions.

The political sociology approach to policy instruments

Sociology and political science have long been interested in the analysis of policy instruments. Linder and Peters (1998) attributed the initial questions to American scholars (Dahl and Lindblom 1953) and, at a later stage, to the American institutional and neo-institutional literature (Hall 1993). Significant studies have also been carried out outside the United States, in countries such as the United Kingdom (Hood 1983) and the Netherlands (Kickert, Klijn, and Koppenjan 1997).

Various approaches can be identified in the field (see Hood 1983 for a review). Among them, that of political sociology, on which this chapter builds, aimed to feed the debate on public action. It emerged from an increasing awareness, as highlighted by empirical research, that instruments of public action represent a crucial variable to explain actors' dynamics and policy innovation. The contributions that appeared in the volume *Gouverner par les instruments* (Lascoumes and Le Galès, 2004 – translated into English in 2007) focused precisely on this. Their goal was to move beyond the study of actors, ideas and institutions to reach out to the technologies of government while reviving classic questions on how collective action is organised.

For the sake of clarity, it should be specified that political sociology scholars intend instruments to be

> a device that is both technical and social, that organises specific social relations between the state and those it is addressed to, according to the representations and meanings it carries . . . with the generic purpose of carrying a concrete concept of the relationship between politics and society.
>
> (ibid., 3)

Political sociologists consider instruments as sociological institutions with a cognitive and normative role: they are understood as pragmatic devices that confront actors with structures of opportunity and influence how they behave (Kassim and Le Galès 2010; Halpern et al. 2014) as they encapsulate values and specific interpretations of the social world and of the mode to regulate it (Lascoumes and Le Galès 2007). They are also interpreted as bearers of a symbolic function in the way they can make legitimate power manifest, and of an axiological one to the extent they state the interests granted by public authorities.

According to Kassim and Le Galès (2010) there are two main strengths of the political sociology approach to policy instruments: first, it can reveal existing and evolving dynamics between the governing and the governed, thus highlighting power dynamics. Second, it acknowledges and investigates their un-neutral nature and the (more or less intentional) effects they can produce, regardless of their set objectives. Menon and Sedelmeier (2010) criticise the fact that this approach 'treats' instruments, and their effects, as part of a deliberate policy strategy. On the contrary, Halpern et al. (2014) insist that its logic is probabilistic: policy instruments are recognised on the one side as carrying potential effects, and on the other as potentially impacted upon by uses and forms of resistance which limit their application or else exacerbate their effects. In other words, the political sociology approach is concerned with linking the elaboration and choice of instruments to their implementation, use and effects in the medium-long term. By doing so it encourages scholars to move beyond the division between politics and policies by exploring their interlinked dynamics. Through this perspective, political sociology scholars try and avoid the reification of the limits associated with neo-institutional forms of functionalism which focus on instruments per se and of the limits typical of constructivism, which concentrates on the knowledge mobilised and incorporated in dispositives (ibid.). Instruments are in fact analysed as an intermediary variable capable of structuring the mechanisms and processes characterising public actions (Fourot 2013) and as an explicative element capable of revealing logics of politicisation and de-politicisation.

Instruments, in a political sociology approach, can contribute to stabilise a problem or to the emergence of an 'espace de sens' and of collective action, in relation to an emerging policy issue (Laurent 2014). In certain cases, the choice and use of instruments can point to the attempts to restrict the public debate and to hide policy issues at stake (John 2011), thus privileging the claims and interests of certain social groups over others. At the same time, it can provide fundamental hints to understand policy change processes and to figure out whether a specific device represents a veritable form of innovation or not (Lascoumes and Simard 2011).

It should be underlined that the political sociology approach has benefited from cross-contamination with various disciplines. As suggested by Halpern et al. (2014), a crucial contribution has come from interaction with political scientists such as Hood (1983), Salamon (2002) and Varone (1998), who, in their theoretical and empirical efforts, have contributed to the analysis of power exercise and of the reconfiguration of modes and scales of political regulation. In this sense, attention to instruments has prompted political sociologists to re-actualise questions of domination and of transformation in the forms of power (Le Galès and Scott 2008), including, for example, the proliferation of new modes of control (Hood 2007; Lascoumes and Simard 2011). However, while overlooking reflections on politics in itself, which has been criticised by some authors as a limit (Hassenteufel 2012; Duran 2010), political sociologists have been articulating their work on the dynamics of public action in order to show the workings of the political scene in which actors play (Leca 2012) and the way they co-ordinate themselves.

Methodology

The empirical data draws from part of the qualitative material collected for a research project on immigrants' entrepreneurial activities and local conflicts in Italy, undertaken in the period 2007–2011. The chapter builds specifically on policy documents aimed at the regulation of phone centre shops: the official proceedings of the relevant public sessions of the Veneto Regional Authority, which led to the approval of Regional Law 32/2007 for the regulation of phone centre shops; and the text of Ordinance 19 approved by the Municipality of Verona in 2009.

All the material was downloaded from the respective institutional websites and analysed by means of a discourse and content analysis. Adding to these documents, narrative evidence has been used by drawing from more than 50 interviews with phone centre operators, phone centre customers and other institutional and non-institutional actors that were conducted to inform the wider research project.

It should be specified that the chapter has no normative intention: it does not seek to identify nor promote 'better instruments' (Peters and Van Nispen 1998 cit. ibid.). Its objective is rather that of providing a critical example of how the political sociology perspective on policy instruments can contribute to debates on urban politics.

Empirical findings

Verona as a case study: the political background of the city and the politicisation of the urban safety policy agenda across Italy

While the focus on the specific case study of Verona does not allow for any generalisation, however, it is arguably relevant to the extent it does exemplify trends that have invested the entire Italian peninsula while raising crucial questions for the study of urban politics.

Verona is a medium-sized town located in the north-eastern part of Italy. It has approximately 250,000 inhabitants, with an incidence of the immigrant population corresponding to 12.7 per cent.[1] It was governed by Christian Democrats for over 40 years. In 1999, a centre-right Forza Italia coalition won the election (58 per cent of votes).[2] During the following mandate, with a centre-left Ulivo coalition (2002–2007), the Lega Nord, a then-regional movement with an ethnocentric ideological matrix that later became a political party, started gaining consensus. The then-head of the regional authority health department was presented as its candidate for the 2007 elections and he was very successful (60.69 per cent of votes). The new coalition built its electoral campaign, and the programme which followed, by insisting they intended to prioritise urban safety issues. They did so by implementing various spatial control initiatives, clearly characterised by an anti-marginal social group approach, aimed at 'reconquering' pockets of urban space that, in their opinion, had been left uncontrolled. Examples include: the restriction of the homeless food service delivery to specific areas out of the town centre; the banning of drinks and food consumption in specific areas of some neighbourhoods; the prohibition of street vendors' activities; and the constant monitoring and regulation of 'ethnic shops', particularly phone centres.[3]

This approach towards urban safety and marginal groups is clearly rather repressive in nature and many scholars have suggested it is most likely to be found among centre-right coalitions. However, Barberis (2009) stressed that approaches on these 'hot' issues do not necessarily differ so much in association with the political colour of local coalitions, if not at a discursive level. In fact, the policy scenario sketched out above must be set in the context of an increasing politicisation of urban safety in the last twenty years.

The process gradually emerged from structural socio-economic changes that went hand in hand with post-industrial transformations, in what can be described as a peculiar political-institutional context (Germain and Poletti 2009), that of Italy, even though it is not unique to it: strategies to 'clean' dangerous *others* from urban space have been largely enacted in other European and American countries, in relation to housing markets, policing and welfare retrenchment.

As far as the specific Italian context is concerned, it should be underlined that until 1989 the definition and management of urban safety issues were the hegemony of the central government. Following the introduction of Law 81/1993, political powers were decentralised and more discretion was granted to mayors. Moreover, Law 125/2008 provided them with wide authority for the enactment of ordinances in the field of urban safety. With sentence 115/2011, the Constitutional Court eventually ruled that ordinances had to be limited in space and time, thus re-defining and limiting mayors' sphere of action. However, many of them have been very active in using these tools to regulate and control urban space. After having been hardly on the agenda of municipal police forces, urban safety instruments started being identified as a priority of local policy-making, particularly during electoral rounds. It should also be pointed out that urban safety issues have been repeatedly addressed in strict conjunction with those relating to pacific cohabitation and immigrants (and other marginal groups), who have been repeatedly 'constructed as enemies' (Maneri 1998; Dal Lago 1998).

Phone centres: the evolution of a booming business

Phone centres are small family-run commercial businesses. The first shops opened at the end of the 1990s. A boom was recorded at the beginning the new millennium as they represented a good business opportunity for immigrants and a chance to secure the renewal of their residence permit, which is strictly connected to proving a regular job (Ambrosini 2009).

Phone centres are mostly managed by male individuals, aged 20 to 35, with different immigrant origins, including Senegal, Morocco and parts of Maghreb. More than 50 could be counted in Verona in 2005. According to the narratives of phone centre operators, only a few remain to date.

Most shops were opened in residential areas and in the urban core, on the ground floor of old (sometimes listed) buildings, often in units previously devoted to local handicraft activities which grew uncompetitive due to the raising of out-of-town shopping malls.

Phone centres draw their business from services offered to customers of immigrant origins. Initially, it was largely telephone and internet access. In spite of the fierce competition of mobile operators and the introduction, as we will see later, of a rigid normative framework, cheap international calls have continued to represent a significant transnational practice. Over time, following increasing requests by customers as well as the need of the sector's operators to ensure the economic sustainability of their activity, new services have been offered, ranging from money transfer to video rental, food and sales, as well as informal and (often) free-of-charge support to fill in bureaucratic documents, such as applications for a residence permit renewal or family reunification. This is how phone centres have transformed into popular meeting spots, arguably due to the lack of adequate alternative meeting spaces for immigrant communities in Verona. And it was in association with this specific nature that conflicts emerged. Indeed, as the 'meeting function' of these shops became more visible, residents and nearby (Italian) shop-owners, particularly those of competing businesses (such as small food stores and newsagents), started raising complaints with reference to noise and forms of decay generated by phone centre customers loitering in and around the shops. As they grew more frequent and stronger, complaints were translated in terms of 'moral panic' (Maneri 2001) connected to anti-social behaviours considered as deviant. In response to them, policy makers at a regional and local level drew up a legislative framework to regulate the sector: Veneto Regional Law 32/2007 and Verona Ordinance 19/2009.

The legislative path of the Veneto Region

Law 32/2007 was signed by the then-head of the regional authority health department and drew from two proposals from the majoritarian centre-right coalition (50.5 per cent of seats), both aimed at regulating the sector of phone centres by addressing: forms of disturbances by phone centres' customers (law proposal Alleanza Nazionale 68/2005); allegations of phone centres as a 'suspect business covering illegal traffic'; phone centres' spatial concentration and the commercial depreciation of the areas they are located in; the lack of a regulatory framework for the sector; and unfair competition to local businesses, with privileged opening hours and no restrictive hygienic/health requirements (law proposal Lega Nord 199/2006).

Analysis of the public sessions that led up to the approval of the regional law[4] clearly highlights that the debate followed two main lines of argument: on the one side, some councillors of the opposing coalition[5] nostalgically recalled the 1960s, when Italians used to go downtown to reach public phones, to stress that phone centres similarly represented an essential public service to be promoted (e.g. opposing coalition councillor Variati, A., Veneto Region minutes of the 93rd Public Session – hereafter identified as: OCC, name, number of public session); on the other side some councillors of the leading coalition strongly supported residents and shopkeepers' complaints.

The second line of argument eventually prevailed, thus adding an urban safety perspective to the discussion (e.g. OCC, Bizzotto, M., 93rd PS). Pressures by voters considerably influenced both the leading and opposing coalitions but they were unprepared to cope with a rather complex and delicate issue. As an OCC stated (Cancian, D., 95th PS): 'From now onwards they will be the only phone centres. It was not exactly what we wanted, but this was the only possible direction to come out of chaos! We could do nothing but somehow regulate the sector'. A few councillors (e.g. Leading Coalition Councillor Bizzotto, M., 95th PS – hereafter identified as: LCC, name, number of public session) also insisted on the shortage of effective instruments for mayors to respond, at a local level, to citizens' complaints, in spite of increasing powers having been devolved to mayors in 2008. What emerges from the proceedings is the impression that no sufficiently comprehensive understanding had been achieved to approach the phenomenon from any alternative perspective but that of safety, combined with a general trend in recent decades to push urban spatial policy debates towards safety and the control of cities through specific ordinances. Arguably, the overarching objective of the discussion was to secure urban space by 'domesticating' (Zukin 1995) phone centres, to offer better conditions for the fruition of its 'public function' (LCC, Cortelazzo, P., 95th PS; OCC, Variati, A., 95th PS) while, at the same time, discouraging customers' aggregation in and around the shop, thus overcoming residents' negative perception (LCC, Cortelazzo, P., 95th PS) particularly with respect to their visibility (OCC, Pettenò, P., 93rd PS).

A series of amendments were put forward, thus introducing further rigidity to phone centre regulation, with respect to that in existence for other businesses (OCC, Frigo, F., 94th PS). In fact, two years after the approval of the regional law, a sentence of the Constitutional Court (25/2009) declared a particularly stringent article illegitimate in the way it entrusted mayors to define shops' localisation – normally a state's responsibility – and it violated EU legislation's right to freely access any means of electronic communication.

The most evident effect of amendments was that of constraining the debate on rational legislative technicalities. A more general discussion in terms of social justice, access to jobs for immigrants and the possible distributive effects of the law did not take place. The normative process was de-politicised (Lascoumes and Le Galès 2004). Arguably, governing coalitions focused discussions on policy tools to reach agreement on methods, since agreement on goals was unlikely given the different views held by leading and opposing councillors. Instruments represented a 'vehicle' for structuring short-term exchanges, leaving aside core content issues (Kassim and Le Galès 2010) that were repeatedly associated with the actual regulation (ibid.). An opportunity was missed to promote policy innovation, starting from the 're-interpretation' of phone centres as a new, atypical type of business. They very much resemble 'bazaars' and had the potential of taking over the function of neighbourhood shops, in a scenario characterised by their progressive disappearance due to the expansion of peripheral shopping malls. This aspect was not taken into consideration at all, in contrast with a more general long-term policy tendency to promote entrepreneurialism in the north-east of Italy (OCC Sernagiotto, R. F. I., 93rd PS) and the alleviation of the bureaucratic burden for local entrepreneurs.

The legislative path of the Municipality of Verona

The political sociology perspective supports the idea that instruments can contribute to the emergence of specific representations of a policy issue. This seems to be the case for the Veronese ordinance too: it did not explicitly define phone centres as public spaces and yet it regulated them as such by setting out restrictions on opening times and the sale and drinking of alcohol. Policy experts dealt with these shops by focusing on the technical aspects of the regulation with the ultimate normative aim to 'control' access to phone centres. As a result, phone centres have undergone a process of normative categorisation, alien to the experiential knowledge of operators and customers (see Tricot 2009), through which specific meeting practices have been transposed into a policy practice of social and political distinction (Lamont and Molnar 2002) for immigrants.

Contrary to what the ordinance itself states, the municipality did engage with consultations only with the most representative employers and labour unions. The normative process unravelled in a web of internal contacts between the mayor, his staff and the director of economic activities, combined with the spatial intervention endeavours of the governing coalition. Phone centres were not consulted. Hence, they made their voice heard through protests, direct confrontation with the staff of the Economic Activities Department (Semprebon 2012) and forms of resistance to police inspections (Semprebon 2013), but did not manage to impact on the mayors' decision. Most phone centre operators are not voters and do not have any form of representation in Verona, so elected politicians did not feel pressurised by their pleas. Many prejudices were held against their activities by residents of Italian origins, to whom elected politicians must respond instead. In this sense, it can be hypothesised that both the regional law and the ordinance had the ultimate symbolic role of making legitimate power manifest and stating the values and interests granted by the regional and local authorities.

In 2010 the Constitutional Court declared Regional Law 32/2007 illegitimate, and nullified the Veronese ordinance as a result. Since then, the regulatory efforts of the local authority have been focusing on kebab shops – another type of shop managed largely by immigrants – which have evolved into meeting spots. With the introduction of the new Historical Centre Plan[6] and later amendments in 2012, no new shop can open in the historical centre, based on the principle of preserving 'urban decorum'.

Conclusions

The main aim of this chapter was to contribute to discussions on urban politics and democracy. It focuses on the interweaving dynamics of policy and politics as entrenched in the phone centres legislative framework. By building on the political sociology approach to policy instruments, it illustrates how specific instruments have been used as a coordination device for actors to reach agreement on methods rather than goals and to avoid debates on actual content issues. In particular, it shows that instruments, at a regional and local level, have acted to constrain discussions within the technicalities of the legislative process and shun concerns about social justice and entrepreneurial promotion.

As hypothesised, it confirms that the adopted spatial interventions are coherent with a trend of increasing politicisation of urban safety which has frustrated innovative policy inputs, thus demonstrating the incapacity of policy makers to approach spatial issues from any perspective but that of safety, in an ongoing effort to respond to voters' anxieties.

Arguably, the legislative process has produced clear-cut boundaries that have contributed to the division of what is 'good' and 'bad' urban (public) space, thus acting against the promotion of social inclusion while leaving a policy issue unresolved: since phone centres have been closing down, their function as meeting spots has been partly overtaken by kebab shops and public parks. Yet the lack of dedicated meeting space is still criticised by many immigrants, thus pointing to a yet-unfulfilled claim to the right to the city.

Acknowledgements

Special thanks to Tommaso Vitale and Roberta Marzorati, who read earlier versions of this chapter and provided important comments.

Notes

1 Statistical data (http://demo.istat.it). Accessed February 20, 2014.
2 Electoral data (http://elezioni.interno.it). Accessed May 20, 2013.
3 Decoro e sicurezza nel Regolamento Polizia Urbana (www.poliziamunicipale.comune.verona. it/nqcontent.cfm?a_id=24185). Accessed February 28, 2014.
4 Public Sessions 93, 94, 95; October to November 2007.
5 It consisted of two right-wing parties and nine left-wing ones, with 42.4 per cent of seats.
6 Norme tecniche per fronti commerciali (http://portale.comune.verona.it/media/_ComVR/Cdr/ Tributi/Allegati/NORME_TECNICHE_FRONTI_COMMERCIALI.pdf). Accessed February 28, 2014.

References

Ambrosini, M. (2009). Le formiche della globalizzazione. In Ambrosini, M. (ed), *Intraprendere tra due mondi. Il transnazionalismo economico degli immigrati*. Bologna: Il Mulino.
Barberis, E. (2009). La dimensione territoriale delle politiche per gli immigrati. In Kazepov, Y. (ed), *La dimensione territoriale delle politiche sociali in Italia*. Roma: Carocci.
Dal Lago, A. (ed) (1998). *Lo Straniero e il Nemico*. Genova: Costa & Nolan.
Duran, P. (2010). *Penser l'action publique*. Paris: LGDJ.
Fourot, A. C. (2013). *L'Intégration des immigrants. Cinquante ans d'action publique locale*. Montréal: Presses de l'Université de Montréal.
Germain, S. and Poletti, C. (2009). Répondre aux mobilisations sociales. Le système policier italien en transition. *Revue Française de Science Politique*, 59(6): 1127–1145.
Halpern, C., Lascoumes, P. and Le Galès, P. (2014). L' instrumentation et ses effets. Débats et mises en perspective théoretiques. In Halpern C., Lascoumes, P. and Le Galès, P. (eds), *L' instrumentation de l' action publique*. Paris: Presses de Sciences Po, 15–59.
Hassenteufel, P. (2012). *Sociologie politique. L' action publique*. Paris: Armand Colin.
Hood, C. (1983). *The Tools of Government*. London: Macmillan.

John, P. (2011). *Making Policy Work*. London: Routledge.

Kassim, H. and Le Galès, P. (2010). Exploring Governance in a Multi-Level Polity: A Policy Instruments Approach. *West European Politics*, 33(1): 1–21.

Kickert, W., Klijn, E. H. and Koppenjan, J. (1997). *Managing Complex Networks*. London: Sage.

Lamont, M. and Molnar, V. (2002). The Study of Boundaries in the Social Science. *Annual Review of Sociology*, 28: 167.

Lascoumes, P. and Le Galès, P. (2004). *Gouverner par les instruments*. Paris: Presses de Sciences Po.

Lascoumes, P. and Le Galès, P. (2007). Understanding Public Policy through its Instruments. *Governance*, 20(1): 1–144.

Lascoumes, P. and Simard, L. (2011). Au prisme de ses instruments: "L'action publique au prisme de ses instruments". *Revue française de science politiques*, 61(1): 5–22.

Laurent, B. (2014). Coopérer pour construire un marché international. Le instruments de la cooperation international et leurs efffets. In Halpern, C., Lascoumes, P. and Le Galès, P. (eds), *L' instrumentation de l' action publique*. Paris: Presses de Sciences Po: 465–492.

Le Galès, P. and Scott, A. (2008). Une révolution bureaucratique britannique. *Revue française de sociologie*, 49(2): 301–330.

Leca, J. (2012). L' Etat entre politics, policies and polity, ou peut-on sortir du triangle des Bermudes? *Gouvernement et Action publique*, 1: 59–82.

Linder, S. H. and Peters, B. G. (1998). The Study of Policy Instruments: Four Schools of Thought. In Peters, B. G. and Van Nispen, F. K. M. (eds), *Public Policy Instruments: Evaluating the Tools of Public Administration*. Cheltenham: Edward Elgar Press.

Maneri, M. (1998). Lo straniero consensuale. La devianza degli immigrati come circolarità di pratiche e discorsi. In Dal Lago, A. (ed), *Lo straniero e il nemico. Materiali per l'etnografia contemporanea*. Genova: Costa & Nolan.

Maneri, M. (2001). Il Panico morale come dispositivo di trasformazione della sicurezza. *Rassegna italiani di sociologia*, XLII(1).

Menon, A. and Sedelmeier, U. (2010). Instruments and Intentionality: Civilian Crisis Management and Enlargement Conditionality in EU Security Policy. *West European Politics*, 31(1): 75–92.

Salamon, L. (ed) (2002). *The Tools of Government*. New York, NY, and Oxford: Oxford University Press.

Semprebon, M. (2012). Urban Conflicts and Immigrants' Engagement. A Comparative Analysis of Two Northern Italian Cities. *Journal of International Migration and Integration*, 14(3): 577–595.

Semprebon, M. (2013). Between Routine Police Checks and "Residual Practices of Expulsion Power": The Impacts of the Anti-Terrorism Law on Phone Centres and the Resistance of Owners. An Italian Ethnography in the "Emergency Season". In Anderson, B., Gibney, M. J. and Paoletti, E. (eds), *The Social, Political and Historical Contours of Deportation*. New York, NY: Springer, 105–121.

Tricot, A. (2009). Vers une écologie urbaine du risque? In Cantelli, F., Roca i Escoda, M., Stavo-Debauge, J. and Pattaroni, L. (eds), *Sensibilité pragmatiques. Enquêter sur l'action publique*. Brussels: P. I. E. Peter Lang, 93–113.

Varone, F. (1998). *Le Choix des instruments des politiques publiques*. Berne: Haupt.

Zukin, S. (1995). *The Cultures of Cities*. Oxford: Blackwell.

13 Negotiating public space in a shopping mall

Pavel Pospěch

Introduction

Public space has been described as a contested space, which is manifested in many different conceptualisations. This chapter concentrates on what has been called 'semi-public space': a space with an ambiguous status, used publicly yet at the same time heavily controlled and influenced by private interests. The review will focus on shopping malls, which have become iconic semi-public spaces both by their media prominence and the academic attention directed to them, and by their popularity in the everyday use of urban populations. The aim of this chapter is to study social control as a phenomenon which contributes to this popularity, identifying three categories of social control and discussing their contribution to the way shopping malls are presented by their management.

In the first part, a general discussion of semi-public space will be undertaken with a very brief review of the present writing on the subject and some theoretical considerations addressed. Next, two key analytical concepts will be introduced: the 'stranger' and 'social control'. With these two terms in mind, three forms of social control in a mall will be introduced: social control by surveillance and exclusion, social control by architecture and social control by normativity. Finally, a conclusion will be drawn, based on the ever-present process of negotiation over the nature of space.

Public and private

First, we need to clarify the terms and the way they are being used here. It is necessary to stop thinking about city spaces in a dichotomic way, as public or private, because doing so requires one-dimensional definitions by means of access and legal ownership, all of which are very disputable. Moreover, the history of western civilization knows very few, if any, cases of ideal 'public space', as described in theory. My suggestion is to consider an imaginary scale, ranging from *private* to *public* (but never really reaching the extremes), with various city spaces placed on it. With such an imaginary scale in mind, we will start with the very meaning of the term 'public'. In doing so, we will limit our scope to the urbanist point of view, in order to dispose of the political and other meanings that tend to be attached to the idea of public space. Thus, a public space will be understood as a city space which is accessible to everyone to such an extent that its accessibility can be enforced by legal means and is not subject to further pre-conditions as is the case with private spaces. This definition presents an ideal which never really existed (Allen 2006; Mitchell 1995), yet, within a scale of ideal types, we can afford to use it. As such, public space cannot be claimed by the private actor, be it an individual or a corporation. Private space, on the other end, would be represented mostly by corporate and residential spaces. Access into such spaces is predetermined by a certain contract, be it a real (economically or legally motivated) or a symbolic one (an emotional contract, such as with family members). To be allowed to enter a certain private space requires being, in fact, a party to such a contract, regardless of its various forms. This view relates to another aspect of private spaces: they tend to be concentrated on a single function rather than multifunctional.

On this imaginary scale, the spaces analysed in this chapter are situated somewhere in the middle. There are different names for them: privatized spaces, quasi-public or semi-public spaces. While all these names have their flaws, I shall refer to them as semi-public. The list includes shopping malls, museums, art galleries, churches, sports stadiums and many other spaces with varying legal, spatial or discursive definitions (Bale 1993; Trondsen 1976). Nevertheless, the one thing which is common to all of them (and the only defining characteristic we can rely on) is that although all these spaces are privately owned (i.e. legally private), the public tends to visit them and appropriate them, or, in other words, the visitors (members of the general public) tend to use these spaces, as if they were in fact public (Friedelbaum 1999). This dialectical relationship between conservation and appropriation constitutes the axis around which the set of meanings attached to semi-public spaces rotate and evolve.

At this point, two key terms need to be introduced to the argumentation: 'social control' and the 'stranger'. Starting from the latter, the stranger is a figure which has traditionally been considered a personification of the 'public' part of city, as thinkers like Simmel (1950), Lofland (1975) or Zukin (1995) have shown. Cities are places where strangers are likely to meet, as Sennett (1977) puts it. An encounter with strangers was what scared urban newcomers in the classic novels of the nineteenth century: the freedom and anonymity brought by a move to the city was accompanied by the frightening experience of urban diversity, the world of strangers. Such diversity must be understood as a permanent source of conflict as it makes urban life exciting on the one hand (because anything can happen), and perilous on the other (because anything can happen) – thus the ambivalence of the figure of the stranger (Simmel 1950). The second conflict is between anonymity and familiarity: the diversity of strangers in public space requires anonymity (Sennett 1977), which is fostered by a set of sophisticated rituals (Goffman 1959). Any intrusion of familiarity, of the personal into the anonymous, is considered threatening for the very nature of public space, as in the case of homeless people, who perform their private acts in public space (Mitchell 1997).

The term 'social control' will, in this text, refer to a set of measures, usually employed by the management of the space, intended to affect the way the visitors act in a way which is in line with the management's interests. There is a difference between the categories of social control and of power. While power seems to be, on most occasions, intentional and motivated, social control, on the other hand, may be quite unintended in its effects: in many cases, certain elements of the design or spatial layout will work towards the goals of the management without the latter being aware of it (or willing to acknowledge it). Such a broad definition of social control has its disadvantages, namely in dealing with 'unmotivated' elements, but it gives us a chance to look at the latent functions of shopping malls.

As has been stated earlier, this chapter will focus on shopping malls. There are two major reasons for this: first, malls seem to be among the most controversial and contested semi-public spaces. Second, research of shopping malls is a well-trodden ground and there are a number of case studies as well as theoretical texts to inform this review. With some simplification, two strands may be identified in the research. One focuses on the 'hard' measures of social control, discussing issues of personal safety and danger (see Davis 1990; Helten and Fischer 2004; Lomell 2004; Matthews et al. 2000; Zukin 1998). This strand discusses at large the processes of securitization and militarization of urban space, tracing back their origins to the criminological concepts of Broken Windows and Crime Prevention through Environmental Design. A second strand focuses more specifically on 'softer' approaches and techniques of social control, which do not necessarily use overt surveillance and physical power to achieve their goals. In analyzing and explaining the nature of social control in shopping malls, these researchers often focus more on the issues of normativity (normality) and expectations and the informal rules of social conduct in general (Allen 2006; Wehrheim 2007).

In this chapter, the focus will be on social control in general and I will attempt to present a basic typology of the kinds of social control in question, connecting the two strands of thought identified above. The discussion will be presented in the form of a review of present research, with three types of social control distinguished: social control by surveillance and exclusion, social control by design and architectural means and, lastly, social control by normativity (normalization and expectations) of social conduct.

The mall

Although social control is aimed at a variety of goals, we have to keep in mind that, ultimately, the means and procedures of social control are based in economic motivation (Wehrheim 2007). In other words, while the existence and implementation of measures such as the mall's house rules gain legitimacy through statements invoking moral values of civility and lawfulness, these rules are in fact instrumental and they are introduced only because 'lawful' behaviour of visitors is, at the same time, profitable for the owners of the space (Shearing and Stenning 1983).

Social control by surveillance and exclusion

These forms of social control have been generally studied in North American and British context tradition (Allen 2006). In general, they refer to the use of CCTV cameras and security guards in an attempt to 'purify' the space, to exclude those who are perceived as a threat to the economic goals of mall management (Zukin 1998). There is a rather straightforward logic behind this: the legal status of mall owners and management allows for a high degree of surveillance, yet it leaves them with virtually no means of punishing the offenders – this must be done by police or state authorities. As a result, social control takes a preventive, exclusive form, often focused on the doorway of the mall – criminologists use the term *actuarial justice* to describe this preventive-minded approach (Franzén 2001). One obvious legitimization of social control and surveillance is the right to private property (Shearing and Stenning 1983), the second is, surprisingly, exclusion itself. This is due to the fact that exclusion is further tied to discourses of safety, health, hygiene and comfort, introduced by the symbolic pastoral power (Foucault 2003) of the mall. Thus, a governing legitimization presents surveillance as a necessary strategy to remove the unhealthy, threatening and disturbing – with the catch that in order to stop the few, the many need to be watched and controlled (Lianos 2003).

All malls have their sets of house rules: lists of unwelcome, unacceptable behaviour, which are displayed by the entrances to the mall. There are differences in the extent to which these are enforced and whether they really represent the norms, as performed in practice. With many shopping malls, it does not seem to be this way: the house rules are only a very basic list and exclusion is based on acts or appearances which are not covered by these rules (Saetnan et al. 2004). The actions prohibited usually include loitering, leaflet distribution, running, skateboarding, cycling or inline skating, begging, sale of magazines such as *The Big Issue*, smoking, taking photographs and others (Siebel and Wehrheim 2003). There is a contradiction, though – as we said at the beginning of this chapter, in order to work social control in a mall must be preventive. However, setting up house rules like this implies a repressive control, which is not really the case. As a result, it is not the actions that social control is based on, but rather the appearances. Thus, when we refer to social control by surveillance and exclusion, we refer to an informal social control. Informal because it is (1) only loosely based on the house rules, (2) uncodified, and, most importantly, (3) subject to the individual judgement of the security operators in charge (Saetnan et al. 2004). This subjective side to the problem gives us a new view on exclusion itself: while exclusion as such is a universal phenomenon, its content – the 'who is actually excluded' – is determined contextually (*in situ*) and individually (*ad personam*). In other words, exclusion in malls is not based on a systematic and/or structurally determined principle.

Still, there is something in common: what all the excluded groups and individuals, be they the homeless, beggars, political activists or skateboarding teenagers, share is some level of non-conformity. This non-conformity is immanent to urban public spaces, with their diversity and strangeness as a basic characteristic of urban life. What happens in the mall, then, is the exclusion of strangers, but since it is an informal, non-systematic exclusion, there is an ongoing negotiation about the definition of 'stranger': a negotiation about belonging and not-belonging. There are no fixed sides in this negotiation and not all arguments are put verbally: it is the presence of the undesirables, their transgressions of place (Dixon et al. 2006) on the one hand and the decisions and exclusive measures on the other, that together constitute a silent dialogue, which we will discuss later.

Social control by architectural means

Architecture, Foucault (1975) has written, is no longer about the external space – its purpose is detailed inner control and visibility of those inside. This could be well applied to the architecture of shopping malls. There are many architectural solutions that contribute to social control in malls, even though some were not originally meant to do so; still, however, they all help to promote the management's vision of the space. A document called *Counter terrorism protective security advice* issued by the British National Counter Terrorism Security Office (NaCTSO 2006) for shopping centres includes, among others, the following suggestions:

> . . . keep public and communal areas – exits, entrances, reception areas, bathrooms – clean and tidy, as well as service corridors and yards . . . [. . .] keep the furniture in such areas to a minimum [. . .] pruning all vegetation and trees, especially near entrances, will assist in surveillance and prevent concealment of any packages.
>
> (NaCTSO 2006, 19)

Architecture gives way to visibility and surveillance. A popular mall layout called 'shopping strip' (Helten and Fischer 2004) is nothing but a cross of two long mall streets (one longer than the other) with shops on the outer sides and a plaza with cafés on the crossing. The obvious advantage of such a layout is visibility: straight streets can be covered with a small number of CCTV cameras and the general absence of corners leaves little spaces for potential deviation to hide in. Furthermore, the street-like design makes control of the crowd easier, as Lehtonen and Mäenpää (1997) have shown.

It is the control of the crowd movement that really matters in architectural solutions. Wehrheim (2007) and Manzo (2005) have revealed the use of various elements for this purpose. These range from physical obstacles (barriers, flower pots and others) to subtle elements, such as placing mirrors (which help to slow the visitors down), posters or exhibitions. Decorative elements are, at the same time, elements of regulation – this includes the general use of colours and lighting, as well as specific elements, like walkways painted on the tiled floors (Wehrheim 2007). Using his American case, Manzo (2005) lists the official goals of mall architecture as follows: allowing a high level of customer flow, ensuring a good visibility of the shops and creating a welcoming atmosphere. The last goal is particularly important for our argument: the atmosphere is intended to be conflict-free (Allen 2006), since potential conflicts make difference and strangeness visible. Two shoppers, who could bump into each other accidentally on a badly designed mall street and would start an argument, would no longer play the roles of customers, which has important consequences for the very definition of the place, as will be shown later.

Two examples can be mentioned here to illustrate the nature of social control by architectural means: the restaurants and the benches. As for the former, restaurants and cafés are typically the most 'public' space in a mall, given the co-presence of mall visitors and its potential to foster different kinds of conduct. As such, the restaurant area is often a subject of particular surveillance by the guards (Manzo 2005). The tables in mall restaurants tend to be rather small with a usual maximum of four chairs – this 'family design' prevents large groups of people sitting together, as this is perceived as a potential source of unrest (Manzo 2005). Also, a large group sitting together is a potential source of behaviour inconsistent with the mall's purpose. Another striking thing is – when compared to a city centre café – the absence of 'corners' – i.e. relatively secluded tables in the corners outside of the main restaurant scene, where it is possible to have a private conversation over a cup of coffee, or just sit and read a book or a paper (Wehrheim 2007). Again, such elements would create a potential for difference, a 'transgression of place.'

The benches and other seating areas present another kind of problem, a typical illustration of the dilemmas the mall management must face. Providing visitors with comfortable benches is an example of good customer care – on the other hand, though, if the benches were too comfortable, the visitors could use them for hanging out with friends, thus spending their time in a non-consuming way. One solution would be removing benches altogether, leaving only the paid seating spaces (cafés, restaurants). However, this strategy can produce the unintended consequence of discouraging certain groups of customers. As a result, there is a general lack of benches in malls (Wehrheim 2007) and those that are installed often tend to be designed in an 'anti-social' way – that is, the seating is arranged in such a way that neighbours' vision angles do not intersect, making communication difficult. One example is a 'tree bench', a short-perimeter circular bench around a tree, where having a conversation means having to look over one's shoulder constantly.

The reason why these two cases – the restaurants and the benches – are important is that they are result of a process of negotiation. The purposeful bench design, as well as the restaurant layout, can both be perceived as the mall management's reaction to the way visitors were appropriating the space: hanging out with friends, informal chatting or reading a paper long after one has finished one's meal – these are undesirable actions, and architectural solutions are one way of dealing with them. Nonetheless, the negotiation process is still ongoing and never really finished: Manzo (2005) reports a group of visitors of Italian origin who regularly meet at 'their' couple of restaurant tables for a friendly talk. While having 'regulars' in their restaurants might be in the end advantageous for the restaurant (and mall) management, it is in their best interest to make sure the restaurant tables are fixed to the floor, so that the visitors cannot hurl them together. This is just an example of a strategy which can be used in the ever-ongoing struggle for the re-appropriation of space.

Social control by normativity (normalization of social conduct)

The third type of social control is the most subtle and difficult to pinpoint. While elements of surveillance and architectural solutions are easily visible, control over social conduct has to be undetectable and not observable, and for this reason there is always danger of prejudice on the part of the researcher. This kind of social control has been emphasized in the works of Siebel and Wehrheim (Siebel and Wehrheim 2003; Wehrheim 2007). The subtlety of this kind of social control is reflected in its measures, as they produce what may be called the 'politics of invisibility'. First, deviance itself is made invisible (Wehrheim 2007), so as not to disturb the harmonic atmosphere of the place: this includes the perpetual cleaning of the place several times a day, or the tendency of the mall management to solve every incivility behind the closed door of the office (as reported by Wehrheim). The principle of invisibility is even more valid for social control, which has a lot to do with the ambiguity of the perception of its elements: while CCTV cameras and guards are, on the one hand, a guarantee of personal safety, on the other hand they also signalize the actual presence of danger. Wakefield (2005) describes two ways of making social control invisible: naturalization (tying social control with other organizational functions) and quaintification (a symbolic re-definition of elements of social control as non-threatening and laudatory). The fact that means of social control are successfully hidden from the visitors has been demonstrated by Wehrheim (2007), who has shown that only a very small number of the visitors actually had an idea about the existence of shopping mall house rules – and those who did know were the same people who had an experience or a potential of being in conflict with them.

While the invisibility of deviance and control certainly contributes towards normalization, we need to explain the core of the 'normalizing' argument. Wehrheim (2007) carried out a comparative survey in German mall and on a central city shopping street, asking passers-by for their opinion what the others around them were doing. While 41 per cent of the mall visitors claimed others were 'just shopping', only 11 per cent of the street respondents thought so. This is a good example of differences in the anticipation of the same kind of behaviour: the others' behaviour in the shopping mall is more easily anticipated, because the others are perceived as homogeneous (Salcedo 2003).

This homogeneity is further linked with familiarity: finding the behaviour of others easy to anticipate means 'knowing' them. Of course, we are not referring to personal knowledge here; the knowledge is based on role expectations: when in a mall, we know what to expect from others and we are able to anticipate their actions, because we suppose that they are the same as us. The homogeneity is not only perceived but also constructed in this way, since if the others behave in the same way as me, it is also me who behaves in the same way as the others. The anonymity, typical for modern urban settings, is breached. Walking through 'traditional' public space, like a city square or a street, we usually know neither the people around us, nor the roles they are playing, which is further magnified by the condition of individualization whereby it becomes increasingly difficult to tell a person by his or her appearance. Walking through a mall, we can guess, expect and anticipate – there is a feeling of safety attached to anticipated clarity. The anonymity and its bearer – the stranger, reduced to a plain consumer, are expelled from shopping malls.

Interestingly, as Lehtonen and Mäenpää (1997) have pointed out, the homogeneity and familiarity is further fostered by the location of the mall: as a large number of malls are built on the outskirts of cities, they present an 'ultimate destination': there is no way beyond the mall; going further means leaving the city altogether. The mall is an end by itself and all who are inside came here to pursue this end. The behavioural pattern of the mall homogeneity is normality (Vaz and Bruno 2003), a socially constructed set of rules which altogether correlate positively to exclusion. Both normality and exclusion are definitions of belonging to the place. The negotiations about them are the ultimate negotiations which take place in a mall: a socially constructed normality becomes reified as a set of informal governing expectations (which then serve as a ground for exclusion) and it also becomes, in a Foucauldian twist, internalized: the norms are further strengthened by the fact that no-one wants to stand outside of them (Vaz and Bruno 2003). Following normality shields us from the unpleasant experience of being picked from the crowd and dragged into the lights of the stage (Wehrheim 2007); just like consumption is an individualizing process, normality is homogenizing (Bauman 2000).

Conclusion

Three broad types of social control have been identified in this review: social control by surveillance and exclusion, social control by architectural means and social control by normalization and social conduct. While in an analysis these types are considered separately, in practice they will typically overlap. From what we have seen, it appears that social control of the mall is a strategy used by the management in the never-ending process of continual negotiation: what is being negotiated are the borders between what is allowed, what is forbidden, what is discouraged and what is excluded – in short, what is at stake is the very definition of the space.

There is a resemblance to the processes of primary socialization whereby a child is taught that 'a street is not a place for ball games' and 'a grocery shop is not where you play hide-and-seek'. Similarly, the negotiation in question may alter the definition of a mall: is it just a place for shopping? Or a place for socializing? For having a little nap on the bench? For celebrating loudly in a mall restaurant?

The management using various means of social control, and the visitors committing everyday incivilities and 'transgressing the place' are the actors of this negotiation process. The 'hard' control measures of securitization are among the most powerful tools that enter the process, intended to exclude the bearers of undesirable conduct and to 'silence their voice' in the negotiation – i.e. not to allow them to pursue their own definitions of the space. It is because their definitions of the space are considered too radical, too dangerous to be considered legitimate: it is a similar kind of exclusion to what Mitchell (1997), as quoted earlier, described for the case of the homeless in public space: the disturbing action *per se* is not the main cause of concern. Rather, threat could potentially alter the dominant definition of space, imposed by the mall management.

What, then, is the product of the negotiation process and where can we find it? We have seen that the negotiated normality serves as a ground for exclusion, for a definition of belonging – which is the purpose of the house rules. Indeed, if the house rules listed all the possible incivilities and transgressions, they would be too restrictive and ineffective. Thus, the socially constructed normality exists as an implicit, shared, yet not verbalized knowledge. It is a tacit knowledge, for it is transferred by experience: rarely does someone tell us what to do and what not to do in a shopping mall or how to behave properly in an art gallery or in a church. We learn the most of this knowledge by experience: the question 'Have you ever been to. . .' relates directly to 'Do you know what. . .is?' Since it is the everyday experience and everyday behaviour where this tacit knowledge is produced, transferred and kept, we must study everyday social conduct in Goffman's terms, the tiny, seemingly banal interactions and negotiations, in order to understand the kind of ambiguous publicness of semi-public spaces.

Acknowledgement

This research has been supported by a Czech Science Foundation (GAČR) grant, No. 14-32200P, 'Incivility in urban public space'.

References

Allen, J. (2006). Ambient Power: Berlin's Potsdamer Platz and the Seductive Logic of Public Spaces. *Urban Studies*, 43: 441–455.
Bale, J. (1993). The Spatial Development of the Modern Stadium. *International Review for the Sociology of Sport*, 28: 121–133.
Bauman, Z. (2000). *Liquid Modernity*. Cambridge: Polity Press.
Davis, M. (1990). *City of Quartz: Excavating the Future in Los Angeles*. London and New York, NY: Verso.
Dixon, J., Levine, M. and McAuley, R. (2006). Locating Impropriety: Street Drinking, Moral Order and the Ideological Dilemma of Public Space. *Political Psychology*, 27(2): 187–206.
Flusty, S. (2001). The Banality of Interdiction: Surveillance, Control and the Displacement of Diversity. *International Journal of Urban and Regional Research*, 25(3): 658–664.
Foucault, M. (1975). *Discipline and Punish: The Birth of the Prison*. New York, NY: Random House.
Foucault, M. (2003). *Myšlení vnějšku* [The Outer Thought]. Prague: Hermann a synove.
Franzén, M. (2001). Urban Order and the Preventive Restructuring of Space: The Operation of Border Controls in Micro Space. *The Sociological Review*, 49(2): 202–218.
Friedelbaum, S. (1999). Private Property, Public Property: Shopping Centers and Expressive Freedom in the States. *Albany Law Review*, 69: 1229–1263.
Goffman, E. (1959). *The Presentation of Self in Everyday Life*. New York, NY: Anchor Books.

Helten, F. and Fischer, B. (2004). Reactive Attention: Video Surveillance in Berlin Shopping Malls. *Surveillance & Society*, 2(3): 323–345.

Lehtonen, T. and Mäenpää, P. (1997). Shopping in the East Centre Mall. In Falk, P. and Campbell, C. (eds), *The Shopping Experience*. London: Sage.

Lianos, M. (2003). Social Control after Foucault. *Surveillance & Society*, 1(3): 412–430.

Lofland, L. (1975). A World of Strangers: Order and Action in Urban Public Space. *Social Forces*, 53(3).

Lomell, H. M. (2004). Targeting the Unwanted: Video Surveillance and Categorical Exclusion in Oslo, Norway. *Surveillance & Society*. 2(3): 346–360.

Manzo, J. (2005). Social Control and the Management of 'Personal' Space in Shopping Malls. *Space and Culture*, 8(1): 83–97.

Matthews, H., Taylor, M., Percy-Smith, B. and Limb, M. (2000). The Unacceptable Flaneur: The Shopping Mall as a Teenage Hangout. *Childhood*, 7(3): 279–294.

Mitchell, D. (1995). The End of Public Space? People's Park, Definitions of the Public, and Democracy. *Annals of the Association of American Geographers*, 85(1): 108–133.

Mitchell, D. (1997). The Annihilation of Space by Law: The Roots and Implications of Anti-Homeless Laws in the United States. *Antipode*, 29(3): 303–335.

NaCTSO (2006). *Counter Terrorism Protective Security Advice for Shopping Centres* www.terrorisminfo.mipt.org/pdf/Counter-Terrorism-Protective-Security-Advice-Shopping-Centres-UK-National-Counter-Terrorism-Security-Office.pdf [accessed July 28, 2008].

Saetnan, A., Lomell, H. and Wiecek, C. (2004). Controlling CCTV in Public Spaces: Is Privacy the (Only) Issue? *Surveillance & Society*, 2(3): 396–414.

Salcedo, R. (2003). When the Global Meets the Local at the Mall. *American Behavioral Scientist*, 46(8): 1084–1103.

Sennett, R. (1977). *The Fall of Public Man*. New York, NY: A. Knopf.

Shearing, C. D. and Stenning, P. (1983). Private Security: Implications for Social Control. *Social Problems*, 30(5): 493–508.

Siebel, W. and Wehrheim, J. (2003). Security and the Urban Public Sphere. *Deutsche Zeitschrift für Kommunalwissenschaften*, 42(1).

Simmel, G. (1950). *The Sociology of Georg Simmel*. New York, NY: Free Press.

Trondsen, N. (1976). Social Control in the Art Museum. *Journal of Contemporary Ethnography*, 5: 105–119.

Vaz, P. and Bruno, F. (2003). Types of Self-Surveillance: From Abnormality to Individuals "at Risk". *Surveillance & Society*, 1(3): 272–291.

Wakefield, A. (2005). The Public Surveillance Function of Private Security. *Surveillance & Society*, 2(4): 529–545.

Wehrheim, J. (2007). Die Ordnung der Mall. In Wehrheim, J. (ed), *Shopping Malls: Soziologische Betrachtungen eines neuen Raumtyps*. Wiesbaden: VS Verlag.

Zukin, S. (1995). *The Cultures of Cities*. Oxford and Cambridge, MA: Blackwell.

Zukin, S. (1998). Urban Lifestyles: Diversity and Standardisation in Spaces of Consumption. *Urban Studies*, 35(5–6): 825–839.

Conclusion
Rediscovering public space globally

Svetlana Hristova and Mariusz Czepczyński

Contemporary public space has undergone radical transformation during the last decade, a result of intense global production. It is conceptualized in this volume as simultaneous processes of occupation and reimagination reflecting diverse contradictory trends: global crises' repercussions in different parts of the world and the mundane tensions and co-operation in the daily routine of citizens; the trade-offs between political influence, state power and private ownership, but also the re-appropriation of public meaning and the reclaiming of social rights by the new poors, the urban precariats. The highlighted cases from Europe, Asia, Africa and North America reveal diverse stories not only of how public spaces are hybridized into new global locales of high-status consumerism, but also how neglected places in these cities are regenerated into green spaces of emancipating cultural alternatives.

Public space has always been a primary public good. However, today it is endangered by the loss of a public sense, crucified between touristification and terroristification – either reduced to the representational realm of watchable/eatable/experience-able touristic pleasures, or transgressed in violation and terror. Still, public space – as the examples in the book testify – even after being conceded to the insatiable consumerism of late modern society, and after the global crises of 2008, can still regain the moral weight that seemingly has been lost.

The two leading narratives in the book, occupation and reimagination, are internally connected and mutually reinforcing. The connotations of *occupation* used in different chapters vary from the mere settling down and accommodation of urban newcomers into a city's life to illegal acts of temporarily taking possession of certain urban places for uses unforeseen by public city regulations (e.g. squatting or guerrilla gardening). *Occupation* as an usurpation and appropriation of places has acquired globally a specific new meaning as an urban tactic of protest, based on space: that of *encamping* as political and cultural resistance, spread from Tahrir in Cairo to Zuccotti Park in New York, Puerta del Sol in Madrid and Paternoster Square in London. The other trend, of reimagination, the change in the design and use of a public place, is also about social imagery which is imposed on the space. Reimagining, then, is the reclaiming of space via the means of public arts, architecture and the 'aesthetics of informality' (Amin and Thrift 2004, 234), constituting new social solidarities and alternatives to the mainstream consumerist culture ranging from emancipatory urbanism to 'civic hedonism'. Finally, the visual transformations in public space are always about power (its enforcement, redistribution and trade-offs), because as underlined by Jerome Krase in the current volume, those who have the power to command and produce space are therefore able to reproduce and enhance their own power.

However, no matter how different are the cases presented in this volume – from the European East and West, from Russia, Turkey, Egypt and the United States, they give evidence of ubiquitous trends reflecting universal changes in contemporary public space, its uses and users:

1 Although public space – since the times of the ancient agorae and fora – has been a premise of social inclusion and more open decision-making (i.e. democratic rule as opposed to autocracy), never before has the public awareness about its importance been so high, and never before has it been so instrumentalized: public space today has become a tool of both urban politics and social unrest; an instrument for calming down the society through aestheticization and different systems of observation and control, but also a platform enabling the expression of social claims and public concerns – such as those of the global risk society longing for security and humanization.

2 There is a constant tension between the growing discontent and social upsurge which found an outlet in public spaces, on the one hand, and their pacification through different forms of social control on the other. The diversification and evolution in the conventions of regulation, surveillance and control range from mild forms of aesthetic upgrading, touristification, and domestication by Cappuccino (Zukin 1998) to harder police control and disciplination as a response to social aggravation and unrest in public spaces which ultimately lead to the disappearance of otherness (Pospěch, current volume). Terrorist acts are the most radical negative form of contestation, and currently turn public spaces of major cities in Europe and the United States into a battleground of a new anti-Western war.

3 In this struggle over public space, a higher degree of public consciousness has developed concerning design, uses, and access. Higher expectations and confrontations have also arisen regarding how we want to see our urban commons. The class/place debates have been exercised in many squares, streets and parks for decades around the Western world. In the last ten years this process has intensified, especially in Central and Eastern European post-socialist cities (Czepczyński, current volume).

4 In public space as the domain of the visual, forms of control can also be tools for cultural resistance, 'the minor, everyday initiatives and associations [. . .] that develop emancipatory content within existing political structures and initiate social and cultural change' (Amin and Thrift 2004, 234).

5 Although the consumerist spirit of contemporary culture is still very strong and continues to pervade the urban scene, the global crisis made it obvious that insatiable consumption is destructive for societies and a fuel in the engine of non-sustainable development. A new ethos of post-production began to develop oriented to the satisfaction of human needs, but against conspicuous consumption. Consequently, consumption in all its material and symbolic forms will determine less the content of public space activities, as long as consumption 'does not preclude encouragement of diverse behaviours and the creation of relaxed mixed environments' (Carmona 2015, 391).

6 The increasing hybridization of public space – of its ownership, management and uses – reveals the growing dependence of the public sphere on private capital (Zukin, current volume). In post-socialist countries, such hybrid spaces with

high representative value aim to adopt a 'global' outlook with a visible emphasis on youth, Westernization, commercialism, and cleanliness (Dixon, current volume). The hybridization is also expressed in mixed 'participatory' forms of city governance. This suggests that such integration of social movements into enduring bureaucratic structures weakens their utopian content (Rode and Schwab, current volume), or that formal administrative decisions can entirely terminate promising ideas for more human public spaces (Semprebon, Garcia Jerez, current volume).

7   Globalization, including vast migration processes, led to unprecedented diversification of publics present in the open spaces of modern cities, and the appearance of the 'transnational figure' in the city – immigrants, travelling professionals, creative-class bohemians, all those post-modern nomads, to use Bauman's metaphor, who live by moving around the globe. Furthermore, the born-by-the-crisis new poors turned into precariats, and although deprived of their houses and savings, possess cultural and social capital, which has been enough to mobilize on a mass scale. Their knowledgeable discontent did not just pour out into the public spaces of the modern world, but was capable of moving fluently between these spaces, transmitting information, new visions and practices concerning urban alternatives (Hristova, current volume).

8   This marks the process of the transnationalization of public space, with new spatial tactics of globally connected locally-based social actors. The various forms of emancipatory urbanism comprehend 'the interplay between local initiatives and global discourses as a potential renewal for forms of contestation' (Mayer and Boudreau 2012, 278).

9   The transnationalization of public space would be impossible without the agency of social media. By the end of the twentieth century we saw the development of a new public sphere connecting into an inseparable 'third reality,' the virtual and real space as a global-local continuum.

10 The changes discussed in this volume necessitate a rethinking and redefinition of contemporary public space by taking into consideration all its constituent elements – ownership is just one of them. Private ownership, or public-private governance (e.g. the BID model that has spread into many Western cities), is articulated differently on a local level (Michel and Stein 2015). And, in spite of the neoliberal threat of creeping privatization of public resources, they still can produce functioning public spaces (Zukin, current volume; Carmona 2015). Moreover, in the European context, BIDs can even be 'safeguards of public space rather than jeopardizing it' (Michel and Stein 2015, 79). The diversity of publics, the diversity of their activities and the meanings invested in public space are other criteria for its open and free constitution. To put it simply, as Richard Sennett did, 'the most important fact about the public realm is what happens in it' (Sennett, no date appointed). In the spirit of Habermas, what happens in public space depends not only on the ownership and governance arrangements, nor on the institutional guarantees of the constitutional state; genuine public space 'also needs the supportive spirit of cultural traditions and patterns of socialization, of the political culture, of a populace accustomed to freedom' (Habermas 1992, 453).

References

Amin, A. and Thrift, N. (2004). The 'Emancipatory' City? In Lees, L. (ed), *The Emancipatory City? Paradoxes and Possibilities*. Thousand Oaks, CA: Sage.

Carmona, M. (2015). Re-theorising Contemporary Public Space: A New Narrative and a New Normative. *Journal of Urbanism: International Research on Placemaking and Urban Sustainability*, 8(4), 373–405.

Habermas, Y. (1992). Further Reflections on the Public Sphere. In Calhoun, C. (ed), *Habermas and the Public Sphere*. Cambridge, MA: MIT Press.

Mayer, M. and Boudreau, J-A. (2012). Social Movements in Urban Politics: Trends in Research and Practice. In Mossberger, K., Clarke, S. E. and John, P. (eds), *Oxford Handbook on Urban Politics*. Oxford: Oxford University Press, 208–224.

Michel, B. and Stein, C. (2015). Reclaiming the European City and Lobbying for Privilege: Business Improvement Districts in Germany. *Urban Affairs Review*, 51(1): 74–98.

Sennett, R., *The Public Realm*. Available at: www.richardsennett.com/site/SENN/Templates/General2.aspx?pageid=16 [accessed July 10, 2016].

Zukin, S. (1998). Politics and Aesthetics of Public Space: The 'American' model. *Ciutat real, ciutat ideal. Significat i funció a l'espai urbà modern* [*Real city, ideal city. Signification and function in modern space*]. Barcelona: Centre de Cultura Contemporània de Barcelona.

Index